D0018707

DO
YOU
BELIEVE
IN
MAGIC?

ALSO BY PAUL A. OFFIT, M.D.

Deadly Choices: How the Anti-Vaccine Movement Threatens Us All

Autism's False Prophets: Bad Science, Risky Medicine, and the Search for a Cure

Vaccinated: One Man's Quest to Defeat the World's Deadliest Diseases

The Cutter Incident: How America's First Polio Vaccine Led to Today's Growing Vaccine Crisis

DO YOU YOU BELIEVE IN MAGIC?

Vitamins, Supplements, and All Things Natural: A Look Behind the Curtain

PAUL A. OFFIT, M.D.

HARPER

NEW YORK · LONDON · TORONTO · SYDNEY

HARPER

A hardcover edition of this book was published in 2013 by HarperCollins Publishers.

FIRST HARPER PAPERBACK PUBLISHED 2014.

Designed by William Ruoto

Library of Congress Cataloging-in-Publication Data has been applied for.

ISBN 978-0-06-222298-5 (pbk.)

20   OV/LSC 10 9 8 7 6 5

To all the science writers, science advocates,
and science bloggers who have dared
proclaim that the emperors of pseudoscience
have no clothes

When religion was strong and science weak,
men mistook magic for medicine.
Now, when science is strong and religion weak,
men mistake medicine for magic.

—THOMAS SZASZ

Contents

PROLOGUE
Taking a Look at Alternative Medicine

Americans love alternative medicine. They go to their acupuncturist or chiropractor or naturopath to relieve pain. They take ginkgo for memory or homeopathic remedies for the flu or megavitamins for energy or Chinese herbs for potency or Indian spices to boost their immune systems. Fifty percent of Americans use some form of alternative medicine; 10 percent use it on their children. It's a $34-billion-a-year business. My friends are no different. One uses cold laser therapy for his allergies, another takes a homeopathic remedy named oscillococcinum to cure her colds, and a third swears that acupuncture is the only thing that relieves his back pain.

Furthermore, alternative medicine—which in the 1960s was denigrated as fringe or unconventional medicine—has entered the mainstream. Hospitals have dietary supplements on their formularies or offer Reiki masters to cancer patients or teach medical students how to manipulate healing energies. In 2010, a survey of six thousand hospitals found that 42 percent offered some form of alternative therapies. When asked why, almost all responded, "patient demand." Big Pharma is also jumping in. On February 27, 2012, Pfizer acquired Alacer Corporation, one of the country's largest manufacturers of megavitamins.

The reason alternative therapies are popular is simple. Mainstream doctors are perceived as uncaring and dictatorial, offering unnatural remedies with intolerable side effects. Alternative healers, on the other hand, provide natural remedies instead of artificial ones, comfort instead of distance, and individual attention instead of take-a-number-and-wait-your-turn inattention.

Like many people who have spent time in today's health-care system, my experiences have been largely disappointing.

I was born with clubfeet. Within hours, both feet were put in casts; the left foot healed; the right didn't. When I was five years old, a surgical procedure was performed on my right foot; one of the first of its kind, my case was later written up in a medical journal. The good news is that my right foot no longer turns awkwardly down and inward. The bad news is that walking is always somewhat painful for me.

While in medical school, I volunteered for a twenty-five-mile walkathon for the National Multiple Sclerosis Society. After completing the walk, the pain in my foot was so bad I had to use crutches for a few days. I visited an orthopedist, who told me I had severe osteoarthritis and that my X-ray looked like that of a seventy-year-old man. I was twenty-four. For most of my adult life, I've tried conventional nonnarcotic pain medicines without relief.

When I was in my thirties, I noticed a small dark spot—no bigger than the head of a pin—on the front of my nose. I ignored it. Twelve years later, my wife suggested I have it removed. The procedure was fast and painless. But a few days later, the dermatologist called with some bad news. He had received a report from the pathologist. The diagnosis: metastatic malignant melanoma. A death sentence.

I panicked and immediately called the pathologist. "This diagnosis doesn't make any sense," I pleaded. "How could I have a metastatic lesion on only one part of my body that has remained unchanged for more than a decade? And where's the primary cancer, the place from which the metastasis had supposedly spread? Doesn't this make me the longest-living survivor of untreated metastatic melanoma in history?!" The pathologist was sympathetic but unfazed. The diagnosis was what it was. If I wanted her to, however, she was willing to send my biopsy to the nation's foremost expert on melanoma: a dermatopathologist in New York City. A few weeks later, he called with his diagnosis: metastatic malignant melanoma. He patiently explained that, given where the malignant cells were and what they looked like, it couldn't be anything else.

For the next two years, I went to the dermatology clinic at the University of Pennsylvania, getting periodic physical examinations, chest X-rays, and blood tests looking for evidence of further metastases. None were found. Also, no one could find the original site from which my melanoma had supposedly spread. A mystery, they claimed.

Later, my wife, who is also a doctor, sent my biopsy to a dermatologist friend of hers, who said that I didn't have malignant melanoma—my real diagnosis was cutaneous blue nevus syndrome, a benign disorder that mimics melanoma. I was happy to be done with it. But two years of thinking that I was suffering from a fatal illness had been hell.

When I was in my early fifties, a sharp, persistent pain in my left knee made it difficult to walk. Unable to tolerate it any longer, I visited an orthopedist, who diagnosed a partially torn medial meniscus (the cartilage in the knee that keeps bone from rubbing against bone). The surgery will be simple, he explained, with a full recovery in a few days. But in the postoperative haze of anesthesia, I learned that it hadn't been that easy. The orthopedist explained that my problem wasn't a torn meniscus after all; it was a loss of cartilage behind my kneecap. Instead of minor knee surgery, I had just undergone microfracture surgery, where small holes are drilled into bone. The recovery wasn't going to be a few days—it was going to be a year. The miscalculation didn't seem to surprise or upset the orthopedist. But it upset me.

By my mid-fifties, consistent with my age, I began to suffer symptoms of an enlarged prostate. Now I was in the world of urologists, which meant I would periodically get my PSA level checked. PSA, or prostate-specific antigen, is supposedly a predictor of prostate cancer. But the more I read studies about PSA, the more I realized it isn't a very good predictor at all. Even biopsies of the prostate are confusing. As it turns out, most men with prostate cancer die *with* the cancer, not *from* it. Which means that most men with prostate cancer have needless surgery. And the surgery is brutal, leaving many incontinent and impotent. As a consequence, urologists have varying opinions about how to avoid prostate cancer.

During these misadventures, I've gotten a lot of advice from a lot of people. Some have gone as far as to suggest I abandon conventional medicine. They said I should take saw palmetto for my prostate and chondroitin sulfate and glucosamine for my foot and knee pain—all readily available without a prescription. They told me that I shouldn't have seen an orthopedist—I should have seen an acupuncturist or a chiropractor—and that I shouldn't have gone to a urologist for prescription drugs: I should have gone to a naturopath for something more organic, more natural. They urged me to stop being so trusting of modern medicine and to once and for all take control of my health—to leave a system that was clearly flawed.

So I went to the General Nutrition Center and bought saw palmetto, chondroitin sulfate, and glucosamine. But before I took them, I looked to see whether studies had been done showing they worked. The studies were large, internally consistent, well controlled, and rigorously performed. And the results were clear: saw palmetto didn't shrink prostates, and chondroitin sulfate and glucosamine didn't treat joint pain. Then I reviewed studies of acupuncture, naturopathy, homeopathy, and megavitamins, which also showed results far less amazing than my friends had led me to expect. Some therapies worked; most didn't. And for those that did work, it was *how* they worked that was surprising.

Perhaps most concerning, I found that alternative therapies could be quite harmful. Chiropractic manipulations have torn arteries, causing permanent paralysis; acupuncture needles have caused serious viral infections or ended up in lungs, livers, or hearts; dietary supplements have caused bleeding, psychosis, liver dysfunction, heart arrhythmias, seizures, and brain swelling; and some megavitamins have been found to actually *increase* the risk of cancer. My experience wasn't limited to reading medical journals. As head of the therapeutic standards committee at our hospital, I learned of one child who suffered severe pancreatitis after taking more than ninety different dietary supplements and another whose parents insisted on using an alternative cancer cure made from human urine.

What I learned in all of this was that, although conventional therapies can be disappointing, alternative therapies shouldn't be given a free pass. I learned that all therapies should be held to the same high standard of proof; otherwise we'll continue to be hoodwinked by healers who ask us to believe in them rather than in the science that fails to support their claims. And it'll happen when we're most vulnerable, most willing to spend whatever it takes for the promise of a cure.

The purpose of this book is to take a critical look at the field of alternative medicine—to separate fact from myth. Because the truth is, there's no such thing as conventional or alternative or complementary or integrative or holistic medicine. There's only medicine that works and medicine that doesn't. And the best way to sort it out is by carefully evaluating scientific studies—not by visiting Internet chat rooms, reading magazine articles, or talking to friends.

INTRODUCTION
Saving Joey Hofbauer

They were small
And could not hope for help and no help came.
—W. H. Auden, "The Shield of Achilles"

My first exposure to alternative medicine came by way of a story that circulated during my pediatric residency in the late 1970s. It involved a popular alternative cancer remedy called laetrile. Some might read what follows and feel assured it could never happen today—that no parent would ever do such a thing. But every single influence that drove these parents to do what they did is still very much alive, arguably even more so than it was then.

The story concerns a little boy from upstate New York.

On October 5, 1977, Joey Hofbauer complained to his mother about a lump on his neck. When the lump didn't go away, she took him to their family doctor, Denis Chagnon, who prescribed penicillin, without effect. When the lump got bigger, Chagnon referred Joey to an ear, nose, and throat specialist, Dr. Arthur Cohn, who, on October 25, biopsied it at St. Peter's Hospital, in Albany. Two days later, Cohn had his diagnosis: Hodgkin's disease, a cancer of the lymph glands. Joey was seven years old.

Although the news was devastating, Joey's prognosis was excellent. By the early 1970s, investigators had proved that radiation and chemotherapy offered Joey a 95 percent chance of recovery—with proper treatment, Joey could live a long and fruitful life. But for Joey Hofbauer, the road to recovery wasn't going to be easy. Within weeks, a battle erupted over how Joey should be treated and by whom. On one side were Joey's parents, citizen activists, the media, the John Birch Society, and a movie star. On the other were cancer specialists, Senator Edward Kennedy, the Saratoga County Department of Social Services, and the Food and Drug Administration (FDA). The battle lasted three years.

When he learned that Joey had Hodgkin's disease, Arthur Cohn advised the Hofbauers to see a cancer specialist. The specialist would determine the extent of Joey's cancer by taking biopsies of the liver and spleen. Then Joey would receive radiation and chemotherapy—medicines like procarbazine, prednisone, vincristine, and nitrogen mustard. Cohn reassured the Hofbauers that their son had an excellent chance of survival. But John and Mary Hofbauer weren't reassured. They heard words like *radiation* and *chemotherapy*, and it scared them, conjuring up images of hair loss, vomiting, diarrhea, anemia, and worse. Certainly there was a better way to treat their son—a more natural way. So they rejected Cohn's advice and signed Joey out of St. Peter's Hospital. On November 8, the Hofbauers flew their son to the Fairfield Medical Center, in Montego Bay, Jamaica, to receive a remedy they believed was far gentler, far kinder, and far more reasonable than those recommended by Dr. Cohn: laetrile, a natural remedy made from apricot pits.

The day the Hofbauers flew to Jamaica, Denis Chagnon wrote a letter: "Dear Mr. and Mrs. Hofbauer, I have repeatedly asked for the name and address of a physician to whom I can send [Joey's] records. I spoke with Mrs. Hofbauer in the morning hours of Friday, November 4th, and again on Monday, November 7th, and was not provided with an answer. Without treatment [Hodgkin's] disease is oftentimes fatal. I ask you again to provide me with the name and address of his present physician [to] ensure that [Joey] is being properly cared for. If this is not provided by noon, Thursday, November 10th, the following action will be taken: notification of the State Health Department, the Children's Protective Agency, and the American Cancer Society." When the Hofbauers left for Jamaica, Chagnon carried out his threat, reporting them to child services. On November 9, the Department of Social Services of Saratoga County, New York, charged John and Mary Hofbauer with neglect, seeking to remove Joey from the home. The law was clear: "The State, under appropriate circumstances, may provide medical care for a minor where the parent or guardian fails to do so."

On November 23, the Hofbauers returned from Jamaica. Because written and telephone correspondence had been ignored, on November 29, Richard Sheridan and Diana Fenton, from the Department of Social Services—accompanied by an armed sheriff's deputy—visited the Hofbauers. Sheridan remembered what happened next: "[Mr. Hofbauer said] that we weren't going to take his child away unless the Sheriff's deputy drew his gun and arrested him." Sheridan told Hofbauer that the state of New York was now in charge of Joey's care. "I told him there was a [hearing]," recalled Sheridan, "and he said it was illegal because he wasn't there. I said that this was not the place to be talking about this, and Mr. Hofbauer yelled very loudly that it *was* the place to be talking about this, and he wanted everybody to know that we were coming to take his son." Hofbauer was convinced that cancer specialists would only harm Joey. "He said I wanted to take his son and poison [him]," said Sheridan. "He said 'Do you know what chemotherapy is? It's nitrogen mustard gas. It was declared illegal in the wars.'"

When the dust settled, the Hofbauers relented. Through their lawyer, they worked out a deal. Joey would be taken to St. Peter's Hospital with an understanding that no diagnostic tests would be performed and no treatment would be administered—at least not until the case could be heard in family court. Joey stayed at St. Peter's from November 29 to December 9. But John Hofbauer couldn't watch silently while his son was denied what he believed was a lifesaving medicine. So he secretly gave Joey several doses of laetrile until "we were threatened with armed guards at the door, at which time we desisted."

In December 1977, the case of Joey Hofbauer was referred to Saratoga Family Court judge Loren N. Brown, who, much to the dismay of child services, agreed to let the Hofbauers treat their son with laetrile for six months. On one condition:

they had to find a licensed physician willing to do it. "I had a situation where I had to find a doctor in a hurry," recalled Hofbauer, "because everybody was demanding to know who my doctor was." First, Hofbauer asked Dr. Milton Roberts, in Westchester County. But Roberts worried the case had become "too hot to handle," so he turned him down. Then Hofbauer asked Michael Schachter, a psychiatrist from Nyack, New York. Schachter agreed, but only if the Hofbauers signed a consent form releasing him of all responsibility: "I agree to undergo care with Michael B. Schachter, MD," it began. "I understand [that] among the substances, medications, or drugs available [laetrile] may be advised for the purpose of metabolic support. The predominant medical view, including that of the Food and Drug Administration (FDA) and the American Medical Association (AMA) is that this substance [has] no known value [for] the treatment of any disease . . . I understand that some alleged authorities associated with the FDA and AMA assert that the use of this substance constitutes quackery and amounts to a hoax on the American public. I further understand that American physicians have been indicted in California for the use of this substance. I understand that Dr. Michael Schachter [is] not a cancer specialist and [has] no direct experience with the orthodox cancer therapy modalities of chemotherapy, radiation or surgery [and is] not in a position to advise me as to the relative benefits and risks of those treatments for my condition." On December 14, John and Mary Hofbauer signed Michael Schachter's consent form.

Six months later, in June 1978, the court would reconvene to see whether laetrile was working and to determine who would care for Joey Hofbauer: his parents or the state.

Michael Schachter didn't limit his therapy to laetrile. For the next six months, he also gave Joey raw milk, raw liver juice, cod liver oil, soft-boiled eggs, Staphylococcus phage lysate (staph bacteria infected with a virus), pancreatic enzyme enemas (which partially dissolve the lining of the colon), massive doses of vitamin A (which cause blurred vision, bone pain, and dizziness), a vaccine to prevent *"Progenitor cryptocides"* (a bacterium believed by a physician named Virginia Livingston to cause all cancers), a vegetarian diet, daily coffee enemas made by adding three heaping tablespoons of regular coffee to one quart of water (coffee enemas had already caused two deaths), seven injections of an "autogenous vaccine" (made from bacteria in Joey's urine), and Wobe-Mugos enzymes (a combination of several pancreatic enzymes obtained from pigs). None of these therapies had been approved for use in people, and all were arguably in violation of New York State laws on human experimentation. A cancer specialist who later testified at Joey's trial called it "a witch doctor's diet."

In June, six months into Joey's unconventional treatments, the Saratoga County Department of Social Services, Dr. Michael Schachter, and several cancer specialists appeared before Judge Brown to determine whether Joey's alternative cancer cures were working. Most damning was the testimony of John Horton, a professor of medicine at Albany Medical College and a board-certified cancer specialist, who had recently examined Joey. "On feeling the left side of the neck there was a [large] lymph node under the angle of the jaw," he said, "and just below that another [large] lymph node [and] a string of lymph nodes coming down the neck as far as the clavicle [collarbone]." At the time of his diagnosis, Joey Hofbauer had had one swollen lymph gland; now he had seventeen. Dr. Anthony Tartaglia, a board-certified hematologist and chief of medicine at St. Peter's Hospital, had also examined Joey. "There is no question in my mind that the extent of Hodgkin's disease in [Joey] is much greater than when I examined him in December," he said. Tartaglia added that the laetrile that Joey had received was the "equivalent of not getting any treatment."

There were other worrisome signs. Tests showed that Joey had liver damage, most likely caused by dangerously high doses of vitamin A. Also, Schachter apparently didn't realize that Joey's "occasional nausea and abdominal cramps" were probably caused by cyanide poisoning from large doses of laetrile, having never obtained blood cyanide levels to check it out.

Unlike the cancer specialists who had examined Joey, Michael Schachter believed his program was working. "I think he is doing very, very well," he said. "I'm just not as concerned about these lymph nodes in the neck as the other physicians. I feel that [laetrile and metabolic therapy] will be playing a major role in the way medicine is practiced over the next five to ten to fifteen years and consequently I would say that his treatment has been more than adequate, it has been superior." The Hofbauers brought in their own experts—specifically, laetrile promoter Hans Hoefer-Janker; laetrile's inventor, Ernest Krebs Jr.; and Marco Brown, who ran the Fairfield Medical Center, in Jamaica. On July 5, Judge Brown ruled in favor of the parents, stating that they were "concerned and loving" and that Dr. Schachter was "duly licensed."

Although the cancer had spread into his neck, Joey was in the early stages of Hodgkin's disease. And the Saratoga Department of Social Services wasn't giving up. There was still time. Unfortunately, public sentiment was turning in favor of laetrile, making it harder and harder for Joey to get the medicines he needed to save his life.

B y the end of the 1970s, laetrile wasn't just a drug; it was a social movement.

Led by Robert Bradford, of Los Altos, California, the John Birch Society—an ultraconservative organization dedicated to eliminating government regulations—founded the Committee for Freedom of Choice in Cancer Therapy. By 1977 the committee claimed five hundred chapters and thirty-five thousand members. Committee members influenced popular television programs like *60 Minutes*, magazines like *Newsweek*, and commentators like James Kilpatrick, all of whom promoted the wonders of laetrile. Almost singlehandedly, they successfully rallied public support for the drug. In 1976, Alaska became the first state to legalize both the manufacture and sale of laetrile; by 1978 fourteen states had followed; by 1979, twenty-one. Most Americans favored the legalization of laetrile; by 1980 it was a billion-dollar-a-year industry. A movement had been born—a movement that would soon include one of the most popular movie stars of the day.

I n the summer of 1978, Steve McQueen (*The Great Escape*, *The Thomas Crown Affair*, *Bullitt*, *The Towering Inferno*) suffered from a persistent cough and weight loss. Doctors diagnosed him with bronchitis, then walking pneumonia, then a fungal infection. Eventually a lung biopsy revealed the problem: mesothelioma, an aggressive type of lung cancer. After learning he had cancer, McQueen checked into Cedars-Sinai in Los Angeles to begin radiation and chemotherapy, which didn't work. Doctors told him he had only two months to live. So McQueen took matters into his own hands, choosing to treat himself with laetrile at a clinic in Mexico run by William D. Kelley.

Kelley was a flamboyant, charismatic promoter of alternative therapies. Born in Arkansas City, Kansas, he had studied dentistry at Baylor before setting up a clinic in Fort Worth and later in Grapevine, Texas. There Kelley started a mail-order vitamin business. Like Michael Schachter, Kelley believed nonspecific nutritional therapies could treat cancer. Under the direction of Kelley, McQueen received laetrile, massages, shampoos, megavitamins, nutritional supplements, chiropractic adjustments, a high-fiber diet, sheep embryo shots, enzyme implants, and twice-daily coffee enemas (marketed as Kelley's Koffee)—treatments that cost McQueen ten thousand dollars a month (equivalent to eighty thousand dollars today).

Kelley used McQueen's celebrity to promote laetrile. Appearing on the national television show *Tomorrow*, hosted by Tom Snyder, he said, "Those doctors gave him no hope. But his chances are excellent. I believe with all my heart that this approach represents the future of cancer therapy. It took Winston Churchill"—one of the first people to be treated with antibiotics—"to popularize antibiotic medicine. Steve McQueen will do the same for metabolic therapy." McQueen echoed Kelley's enthusiasm; appearing on Mexican television, he said, "Mexico is showing the world this new way of fighting cancer through nonspecific metabolic therapy. Thank you for saving my life. God bless you all."

The John Birch Society's manipulation of the media and the celebrated case of Steve McQueen influenced public opinion. Laetrile had moved into the mainstream. On December 14, 1978, the Saratoga County Department of Social Services appealed Judge Brown's ruling of six months earlier. The case went before Judge Sweeney of New York's Third District Court of Appeals, who reaffirmed the earlier decision: "We are of the view that there is ample proof to support the findings and determination of [Judge Brown's] trial court."

Joey Hofbauer would continue to be treated by Michael Schachter.

The Saratoga County Department of Social Services still had one more appeal—one more chance to save Joey Hofbauer's life. The decision would be made on July 10, 1979. Fortunately for Joey Hofbauer, several events had been set in motion that would soon reduce the public's desire for laetrile. But Joey was getting sicker; the clock was ticking.

On May 26, 1977, Franz Ingelfinger, the distinguished editor of the *New England Journal of Medicine*, published an editorial titled "Laetrilomania." Ingelfinger wrote, "As a cancer patient myself, I would not take Laetrile under any circumstances. If any members of my family had cancer, I would counsel them against it. If I were still in practice, I would not recommend it to my patients." Despite his personal feelings, Ingelfinger suggested a definitive study—one that would settle the argument once and for all. In December 1979, the FDA granted an "investigational new drug" license for laetrile, opening the door for a study. This was the first time in the history of the United States that the FDA had approved human testing of a cancer drug that had never been shown to work in experimental animals.

While researchers were designing Ingelfinger's laetrile study, other events were working on Joey's behalf. In July 1977, Senator Edward Kennedy of Massachusetts held a hearing to discuss the value of laetrile. Testifying in favor of the drug were San Francisco physician and laetrile proponent John Richardson, John Bircher Robert Bradford, and laetrile inventor Ernest Krebs Jr. Kennedy didn't buy it, saying, "There isn't a scintilla of evidence that [laetrile] provides any sense of hope in curing or preventing cancer." During the hearing, representative Terrence McCarthy of Massachusetts was less politic. "The people selling laetrile are crooks, liars, and thieves," he said.

Unfortunately, clear statements by the editor of the *New England Journal of Medicine* and Senator Edward Kennedy didn't convince the courts that Joey Hofbauer had received inadequate care. On July 10, 1979, in response to the Saratoga County Department of Social Services' final appeal, Judge Jasen ruled that he was "unable to conclude, as a matter of law, that Joseph's parents [had] not undertaken reasonable efforts to ensure that acceptable medical treatment is being provided their child." It was Joey Hofbauer's last chance to receive the radiation and chemotherapy he needed. Jasen still considered laetrile, coffee enemas, pancreatic enzymes, and a "vaccine" made from bacteria in Joey's urine to be "acceptable medical treatment."

On July 10, 1980, ten-year-old Joey Hofbauer died of Hodgkin's disease, his lungs riddled with cancer. Although Michael Schachter acknowledged that Hodgkin's disease had killed Joey, he claimed partial success. "Most of the body was either free of Hodgkin's or minimally involved," he said.

Four months later, America's most celebrated standard-bearer for laetrile, Steve McQueen, also died. After McQueen's appearance on Mexican television, Cliff Coleman, a longtime friend, had paid him a visit. "I walked over and there was this skinny old man," recalled Coleman. "No more than a skeleton with dark eyes and a matted beard, sitting swallowed up in an armchair." McQueen told Coleman, "I can't take it anymore." One month later, McQueen was taken to a medical clinic in El Paso, Texas, where tests showed that cancer had spread from his lungs to his abdomen, liver, and pelvis. Within a few days, on November 7, 1980, during surgery to remove a massive abdominal tumor, Steve McQueen died of a heart attack.

One year after the deaths of Joey Hofbauer and Steve McQueen, cancer specialist Charles Moertel, of the Mayo Clinic in Rochester, Minnesota, led research teams at UCLA, the University of Arizona, and the Memorial Sloan-Kettering Cancer Center, in New York, in the clinical trial proposed by Franz Ingelfinger. They treated 178 cancer victims with laetrile and high doses of vitamins, finding that the combination didn't cure, improve, or stabilize cancers. "Patients died rapidly, with a median survival of only 4.8 months," they wrote. "It must be concluded that Laetrile [is] of no substantive value in the treatment of cancers. Further investigation or clinical use of such therapy is not justified." Researchers also found that several patients had suffered symptoms of cyanide poisoning from laetrile. Within a year of the publication, laetrile sales dropped dramatically. In 1987, the FDA banned the sale of laetrile. (It can still be obtained from clinics in Mexico or illegally from the Internet. In recent years, more websites have appeared promoting the drug.)

In retrospect, the last best chance to save Joey Hofbauer had occurred in one court and one court only: Judge Loren Brown's family court. This was the only time that cancer specialists had testified. Lawyers working on behalf of Joey had done their homework. The doctors and scientists presented by the state had published hundreds of papers, written book chapters on Hodgkin's disease, chaired professional societies, headed research teams showing the value of radiation and chemotherapy, performed studies in experimental animals showing that laetrile didn't work and was dangerous, or headed the FDA's section on cancer treatments. They were, in short, the brightest, most accomplished members of their field.

The doctors and scientists offered by the Hofbauers also shared several characteristics: none were board-certified in oncology, hematology, or toxicology; none had ever published a paper in a medical journal; none had shown any reasonable evidence that their therapies worked; and most didn't even have hospital privileges. That Brown could rule in favor of the Hofbauers' choice to deny their son a proven, effective therapy is unconscionable. But an explanation can be found in the record of the trial. In the section titled "Findings of Fact and Conclusion of Law," Brown wrote, "This court finds that metabolic therapy has a place in our society, and, hopefully, its proponents are on the first rung of a ladder that will rid us of all forms of cancer." Brown believed that his small family court in Saratoga County had witnessed a miracle—a breakthrough that would soon turn cancer therapy on its ear. To Judge Brown, the notion that laetrile and coffee enemas could treat Joey Hofbauer wasn't a matter of opinion; it was a "Finding of Fact."

There was another force working against Joey Hofbauer in Judge Brown's courtroom that day—a force far more powerful than clinicians like Michael Schachter or laetrile promoters like Ernest Krebs Jr. or ideologues like Robert Bradford. It was revealed during an exchange between the Hofbauers' lawyer, Kirkpatrick Dilling, and Victor Herbert, a cancer specialist. Dilling was questioning Herbert about the value of bonemeal.

DILLING: Calcium, is that an essential nutrient?

HERBERT: Yes.

DILLING: Are you familiar with the fact that bonemeal is very high in calcium?

HERBERT: I'm familiar with the fact that bonemeal is a dangerous quack remedy because of its lead content and people have died from being given bonemeal instead of calcium properly in milk and milk products.

DILLING: Isn't bonemeal widely available?

HERBERT: Certainly is, your organization pushes it.

Dilling froze. His organization? Herbert had revealed something that wasn't evident to most in the courtroom that day—exactly who was paying for the Hofbauers' defense. Recovering, Dilling went on the offensive. "I want to state for the record," he said, "that I'm proud to represent the National Health Federation and I would appreciate it if the witness would keep his views to himself."

The National Health Federation (NHF) is an organization that represents the financial interests of the alternative medicine industry. At the time of Joey's trial, these therapies had become quite lucrative. Kirkpatrick Dilling was general counsel to the NHF. Against these powerful financial interests, Joey Hofbauer didn't have a chance.

Michael Schachter was never held accountable for his treatment of Joey Hofbauer. On the contrary, since Joey's death Schachter has thrived, directing the Schachter Center for Complementary Medicine, in Suffern, New York. In 2010, a promotional brochure claimed he "has successfully treated thousands of patients using orthomolecular psychiatry, nutritional medicine, chelation therapy for cardiovascular disease, and alternative cancer therapies."

Joey Hofbauer's story, while extreme, contains much of what attracts people to alternative therapies today: a heartfelt distrust of modern medicine (John and Mary Hofbauer didn't believe the advice of hematologists and oncologists); the notion that large doses of vitamins mean better health (Joey was given massive doses of vitamin A, which was likely to have been to his detriment); the belief that natural products are safer than conventional therapies (the Hofbauers preferred laetrile, pancreatic enzymes, coffee enemas, and raw liver juice to radiation and chemotherapy); the lure of healers whose charisma masks their lack of expertise (Michael Schachter, a psychiatrist, convinced the Hofbauers he could cure their son, even though he had no expertise treating cancer); the power of celebrity endorsements (Steve McQueen was one of the most popular movie stars of his day); and, perhaps most of all, the unseen influence of a lucrative business (Kirkpatrick Dilling's NHF, still active today, is one of many lobbying groups that have influenced Congress to offer special protections to the fourteen hundred companies that manufacture alternative remedies in the United States).

Part I

DISTRUST OF
MODERN MEDICINE

1

Rediscovering the Past:

Mehmet Oz and His Superstars

Oh, no, my dear; I'm really a very good man, but I'm a very bad Wizard.

—*The Wizard of Oz*

Few celebrities are more recognizable than Oprah Winfrey. At the height of her syndicated talk show, which attracted more than 40 million viewers a week, Oprah launched the career of a man who would soon become America's most recognized promoter of alternative medicine: Mehmet Oz, star of *The Dr. Oz Show.*

Like Winfrey's, Oz's show is also popular—more than 4 million people watch it every day. It's not hard to figure out why. It's the same reason that John and Mary Hofbauer were attracted to Michael Schachter, or Steve McQueen to William Kelley. Oz believes that modern medicine isn't always to be trusted—that we should retreat to an age when healing was more natural, less cluttered with man-made technologies.

O n the surface, Mehmet Oz would seem to be the last person to argue against modern medicine.

After graduating from Harvard University, the University of Pennsylvania School of Medicine, and the Wharton School, Oz climbed the ranks at Columbia University Medical Center to become a full professor in cardiovascular surgery. He performs as many as 250 operations a year and has authored 400 medical papers and book chapters. Six of his books have been on the *New York Times* best-seller list. Oz was voted one of *Time* magazine's 100 Most Influential People, the World Economic Forum's Global Leader of Tomorrow, Harvard University's 100 Most Influential Alumni, *Esquire*'s Best and Brightest, and *Healthy Living*'s Healer of the Millennium. He's not just famous; he's a brand ("America's Doctor").

Certainly, no one appreciates the advances of modern medicine more than Mehmet Oz. He's a heart surgeon. He holds people's hearts in his hands and fixes them. Oz couldn't do this without anesthesia, antibiotics, sterile technique, and heart-lung machines. But there was one moment when it became clear that Mehmet Oz wasn't a typical heart surgeon. During an operation, "Oz jumped up on a standing stool, peered into the patient's chest, and said, 'I knew we should have used subliminal tapes.'" Oz believed that surgery wasn't enough—success also depended on tapping into his patient's subconscious. Watching this scene was Jery Whitworth, a nurse who operated the heart-lung machine. Whitworth shared Oz's love of alternative therapies. "After a few minutes we stopped," recalled Whitworth, "because the operating room was totally quiet," stunned into silence. Oz, Whitworth, and a group of believers later met secretly to discuss what would eventually become Columbia's Cardiovascular Institute and Complementary Medicine Program. "If the higher-ups had known about these meetings," recalled Whitworth, "they would have disbanded us."

Oz has used his show to promote alternative therapies ranging from naturopathy, homeopathy, acupuncture, therapeutic touch, faith healing, and chiropractic manipulations to communicating with the dead. To understand where Mehmet Oz is coming from, we need to understand where medicine has been.

People have been living on earth for about 250,000 years. For the past 5,000, healers have been trying to heal the sick. For all but the past 200, they haven't been very good at it.

First, people believed disease was a divine act. In Exodus, written around 1400 B.C., God, angry at the Egyptians for their mistreatment of the Hebrews, punishes them with ten plagues, including boils and lice. In Homer's *Iliad*, written around 900 B.C., the god Apollo destroys the Achaean army with a disease ignited by a flaming arrow. In 2 Samuel, written around 500 B.C., God gives David a choice of three punishments for his pridefulness: seven years of famine, three months fleeing his enemies, or three days of plague. David chooses plague, and God obliges, killing 77,000 people. Because God or the gods caused disease, healers were shamans, witches, and priests, and treatments were prayer, amulets, and sacrifices.

Then, starting with the Greek healer Hippocrates in 400 B.C., the focus changed. No longer were diseases defined in supernatural terms; rather, they were caused by something *inside* the body—specifically, an imbalance of bodily fluids called humors. Hippocrates, the father of medicine, named these humors yellow bile, black bile, phlegm, and blood, likening them to four colors (yellow, black, white, and red), four elements (fire, earth, water, and air), four seasons (summer, autumn, winter, and spring), four organs (spleen, gall bladder, lungs, and liver), and four temperaments (choleric, melancholic, phlegmatic, and sanguine). Because diseases were caused by an imbalance of humors, treatments were designed to balance them, most prominently bloodletting, enemas, and emetics (drugs that induce vomiting). Malaria wasn't caused by a parasite; it was the result of excess yellow bile from hot summer weather. Epilepsy wasn't linked to abnormal brain activity; it was caused by too much phlegm blocking the windpipe. Cancer wasn't caused by an uncontrolled growth of cells but by the accumulation of black bile. Inflammation didn't stem from a vigorous immune response; it was caused by too much blood (hence bloodletting).

Two hundred years later, in the second century B.C., Chinese healers embraced a similar concept, reasoning that diseases were caused by an imbalance of energies. Chinese healers treated this imbalance by placing a series of thin needles under the skin (acupuncture). However, because Chinese physicians were prohibited from dissecting human bodies, they didn't know that nerves originated in the spinal cord. In fact, they didn't know what nerves were. Or what the spinal cord was. Or what the brain was. Rather, they interpreted events inside the body based on what they could see outside, like rivers and sunsets. Chinese physicians believed that energy flowed through a series of twelve meridians that ran in longitudinal arcs from head to toe, choosing the number twelve because there are twelve great rivers in China. To release vital energy, which they called *chi*, and restore normal balance between competing energies, which they called *yin* and *yang,* needles were placed under the skin along these meridian lines. The number of acupuncture points—about 360—was determined by the number of days in the year. Depending on the practitioner, needles were inserted up to four inches deep and left in place from a few seconds to a few hours.

And that's pretty much the way things stood until the late 1700s. Practitioners continued to offer therapies based on religious notions of divine intervention or Greek notions of balancing humors or Chinese notions of balancing energies. (Some, such as purgatives, acupuncture, aromatherapy, crystal healing, enemas, magnet therapy, hydrotherapy, and faith healing, are still around today.) But of all the therapies rooted in ancient beliefs, none was more widespread or universally embraced in the eighteenth century than bloodletting. European doctors bled their patients twice a month. Barbers, too, were perfectly willing to bleed their customers. (The red-and-white barber pole represents a white bandage wrapped around a bloody arm.) In the United States, Benjamin Rush, a well-respected Philadelphia physician and signer of the Declaration of Independence, was a big proponent of bloodletting. Rush was so influential that when George Washington suffered epiglottitis (inflammation of the flap of tissue that sits on top of the windpipe), his doctors chose bloodletting instead of the tracheotomy that might have saved his life. Five pints of blood—about half his total blood volume—were taken from Washington as he struggled to breathe. On December 14, 1799, George Washington, a man who had survived smallpox and bullet wounds, went into shock—killed by bloodletting. Sir William Osler, cofounder of Johns Hopkins Hospital, in Baltimore, delivered a fitting postscript. "Man knew little more at the end of the eighteenth century," said Osler, "than the ancient Greeks."

Then medicine took a giant leap forward. Healers no longer believed that illnesses were a matter of spiritual will or humoral imbalances; rather, they defined diseases in biochemical and biophysical terms. This revolution in medical thought centered on several defining moments:

In 1796, Edward Jenner, a country doctor working in southern England, found he could protect people from smallpox by inoculating them with cowpox, a related virus. Jenner's vaccine eliminated smallpox—a disease that had killed as many as 500 million people—from the face of the earth. By inducing the immunity that follows natural infection without having to pay the price of natural infection, vaccines have dramatically reduced deaths from rabies, diphtheria, tetanus, polio, measles, rubella, hepatitis, chickenpox, rotavirus, influenza, yellow fever, typhoid, and meningitis.

In 1854, John Snow, a British physician, investigated an outbreak of cholera in London that had killed more than six hundred people. Snow traced the problem to a water pump on Broad Street. After he removed the pump handle, the outbreak stopped. Snow's observation launched the field of epidemiology and lifesaving sanitation programs.

In 1876, Robert Koch, a German physician, isolated the bacteria that cause anthrax. Knowing that specific bacteria caused specific diseases, scientists could now find ways to treat them.

In 1928, Alexander Fleming, a Scottish biologist, noticed that a mold (*Penicillium notatum*) growing in broth was excreting a substance that killed surrounding bacteria. He called it penicillin. Once-fatal diseases were now treatable.

In 1944, Oswald Avery, an American scientist, found that DNA was the substance from which genes and chromosomes were made, allowing disorders like sickle-cell anemia and cystic fibrosis to be defined in genetic terms.

But it was a relatively unknown Scottish surgeon who—fifty years *before* Jenner's smallpox vaccine—made the single greatest contribution to medical thought. In 1746, James Lind climbed aboard the HMS *Salisbury*, determined to find a cure for scurvy, a disease common among sailors that caused bleeding, anemia, softening of the gums, loss of teeth, kidney failure, seizures, and occasionally death. Lind divided twelve sailors into six groups of two. One pair received a quart of cider every day; the second, twenty-five drops of sulfuric acid three times a day; the third, two spoonfuls of vinegar three times a day; the fourth, a pint of seawater; the fifth, garlic, mustard, radish root, and myrrh gum; and the sixth, two oranges and a lemon. Lind found that only fruits cured scurvy. In 1795, fifty years later, the British Admiralty ordered a daily ration of lime juice for sailors, and scurvy disappeared. (British citizens have been called limeys ever since.)

Although Lind had proved that citrus fruits cured scurvy, he didn't know why. It wasn't until the early 1900s that a

Hungarian biochemist named Albert Szent-Györgyi isolated the substance later called vitamin C, or ascorbic acid (literally, "an acid against scurvy"). Lind's study was groundbreaking because it was the first prospective, controlled experiment ever performed, paving the way for evidence-based medicine. No longer did people have to *believe* in certain therapies; they could test them.

Vaccines, antibiotics, sanitation, purified drinking water, and better hygiene allowed people to live longer. From the beginning to the end of the twentieth century, the life span of Americans had increased by thirty years. None of this increase occurred because healers balanced humors, restored *chi*, or offered sacrifices to the gods; it occurred because we finally understood what caused diseases and how to treat or prevent them.

In a sense, *The Dr. Oz Show* is a voyage back through the history of medicine, starting with our most primitive concept of what caused disease: supernatural forces.

In February 2011, Mehmet Oz asked Dr. Issam Nemeh onto his show. Nemeh is a faith healer. He believes that people can be cured with prayer. One of Nemeh's successes, Cathy, told her story. "I was so sick," she recalls. "I was coughing up blood. I wasn't breathing well. I had a mass in my left lung." Oz showed the audience Cathy's CT scan, which revealed a small, worrisome mass. "I went to see Dr. Nemeh," Cathy continued. "And I had a two-hour visit where we talked and we prayed together. All of a sudden I took this deep breath of air. And I just kept taking breaths. I couldn't believe how much air I was taking in. I felt wonderful." Just like that, Cathy's mass was gone. A second CT scan proved that her lungs were back to normal. No chemotherapy. No radiation. Just prayer. A miracle.

Unfortunately, Cathy's story contained several inconsistencies. First, Oz never mentioned a biopsy, suggesting that the diagnosis had been made by CT scan alone. This should never happen. Because infections can mimic cancer—and because infections are treated differently—a biopsy is required. Second, a closer look at Cathy's CT scan showed that the mass had ragged edges, more consistent with inflammation (seen in bacterial infections) than cancer (where edges are typically smooth). In all likelihood, Cathy had a minor case of bacterial pneumonia that resolved without antibiotics, a common event. Oz's viewers, however, were left with the notion that prayer alone had cured her. (George Bernard Shaw commented on the limits of faith healing after a visit to the shrine of Lourdes. "All those canes, braces, and crutches," he wrote, "and not a single glass eye, wooden leg, or toupee.")

Another example of Oz's embrace of supernatural beliefs can be seen during his surgeries, which look like those of any other surgeon with one exception: the presence of reiki masters like Pamela Miles, a practitioner of therapeutic touch whom Oz has featured on his show. Miles claims that she can detect human energy fields and manipulate them to heal the sick. Oz has never put Miles's claims to the test. But it wouldn't be that hard to do. In fact, it was done a few years ago in a study designed, conducted, and analyzed by Emily Rosa.

Rosa asked twenty-one therapeutic touch healers to sit behind a large partition with two holes at the bottom; she couldn't see them and they couldn't see her. Then she asked the healers to put their hands, palms up, through the holes. After flipping a coin, Rosa put her hand slightly above each healer's right or left hand, asking them to pick which she had chosen. If healers could truly detect her energy field, they would have picked the correct hand 100 percent of the time; if not, about 50 percent of the time. Rosa found that healers were right 44 percent of the time—no different than chance. She concluded, "Their failure to substantiate therapeutic touch's most fundamental claim is unrefuted evidence that [their beliefs] are groundless and that further professional use is unjustified."

In 1999, Emily Rosa published her paper in the *Journal of the American Medical Association*. It was titled "A Close Look at Therapeutic Touch." Unlike Mehmet Oz, Rosa wasn't a cardiovascular surgeon. In fact, she had never graduated from medical school. Or college. Or high school. Or elementary school. When it came time to write her paper, she had asked her mother, a nurse, to help. That's because Emily was only nine years old. Her experiment was part of a fourth-grade science fair project in Fort Collins, Colorado.

Emily didn't win the science fair. "It wasn't a big deal in my classroom," recalled Rosa, who graduated from the University of Colorado at Denver in 2009. "I showed it to a few of my teachers, but they really didn't care, which kind of hurt my feelings." Emily's mother, Linda, recalled that "some of the teachers were getting therapeutic touch during the noon hour.

They didn't recommend it for the district science fair. It just wasn't well received at the school." The press, however, felt differently. Emily appeared on the news on ABC, CBS, NBC, and PBS and was featured in specials by John Stossel, the BBC, Fox, CNN, MSNBC, Nick News, *Scientific American Frontiers*, the Discovery Channel, NPR's *All Things Considered*, the *Today* show, and *I've Got a Secret*. Her story was reported by the Associated Press, United Press International, Reuters, *Time*, and *People* magazine and appeared on the front pages of the *New York Times* and the *Los Angeles Times*. When she was only eleven years old, Rosa spoke at Harvard University in place of the absent Dolores Krieger, the inventor of therapeutic touch and winner of Harvard's tongue-in-cheek Ig Nobel Prize for her claim that human energy fields felt like "warm Jell-O or warm foam." The next day, Emily gave her Harvard speech at MIT. Emily Rosa is listed in *Guinness World Records* as the youngest person to publish a paper in a peer-reviewed medical or scientific journal.

Mehmet Oz's fascination with supernatural forces didn't end with faith healers and therapeutic touch. Later, when he picked John Edward to educate his audience, Oz entered the world of the occult.

Edward is a psychic who communicates with the dead (like the Whoopi Goldberg character in *Ghost*, except without the crystal ball and robes). Oz featured Edward on a show titled "Are Psychics the New Therapists?" "We've had more requests [from our viewers] to join this show than any other we've ever done," gushed Oz. "More than weight loss, more than cancer, more than heart disease. The topic? Do you believe we can talk to the dead?" Oz explained that Edward claimed to have helped thousands of people communicate with loved ones in the afterlife. "A session with a medium can be extremely therapeutic," said Edward.

Oz's interest in the occult came from his experiences in the operating room: "As a heart surgeon, I've seen things about life and death that I can't explain and that science can't address." To Mehmet Oz, John Edward had a gift that was beyond the reach of science. "I want you to know that your mom is okay," Edward told an audience member. "She has a dog with her."

Although Oz promotes Edward's powers, James Randi—a stage magician—doesn't buy it. Randi has appeared on *The Tonight Show Starring Johnny Carson* as well as *Penn & Teller: Bullshit!* In 1986, after receiving the MacArthur Foundation "genius" award, Randi decided to use the money to expose psychics. He now offers $1 million to anyone who can demonstrate clear evidence of paranormal, supernatural, or occult powers. Edward has never taken Randi up on his offer.

According to James Randi, psychics like John Edward employ two basic strategies: "hot reading," which uses information obtained from the audience before the show, and "cold reading," which fishes for information during the show. Randi calls this "hustling the bereaved." It's not hard to see through Edward's claims. When his readings are wrong, Edward claims he has been confused by "energies" emanating from different families. When he has enough wrong guesses, he claims that the "energy" is pulling back. Oz, who is either remarkably trusting, painfully naive, or simply pandering to a gullible public to enhance advertising revenue, never questioned Edward's special gift. "What happens when you start hearing voices," he enthused.

In addition to touting therapies born of the Old Testament notion that supernatural forces caused disease, Mehmet Oz promotes thousand-year-old natural remedies rooted in ancient Greece, China, and India, featuring two men he calls his "Superstars of Alternative Medicine": Andrew Weil and Deepak Chopra, both of whom recommend a variety of therapies (such as acupuncture, plants, herbs, oils, and spices) originally designed to balance humors and restore energies.

Andrew Weil is a balding, white-bearded, slightly overweight man with the demeanor of a guru. A graduate of Harvard Medical School, Weil did an internship at Mount Zion, in San Francisco—a hospital located next to Haight-Ashbury, the epicenter of the hippie counterculture of the 1960s. In the spirit of Ken Kesey (the subject of Tom Wolfe's *The Electric Kool-Aid Acid Test*), Weil fit right in, choosing to study hallucinogenic drugs. In 1972, he published his first book, *The Natural Mind*, in which he claimed that hallucinogens can "unlock" the brain and—in a chapter titled "A Trip to Stonesville"—that "stoned" thinking makes people more insightful. He even celebrated psychosis. "Every psychotic is a potential sage or healer," he wrote. "I am almost tempted to call psychotics the evolutionary vanguard of our species."

After completing one year of a two-year program at the National Institutes of Health, Weil continued to promote his belief that hallucinogenic drugs are good for you. In 1983, he wrote *From Chocolate to Morphine: Everything You Need to Know About Mind-Altering Drugs*. Weil even has a hallucinogenic mushroom, *Psilocybe weilii,* named after him. But Weil's apotheosis came in 1995 with the publication of *Spontaneous Healing,* in which he claimed that health and illness are "manifestations of good and evil, requiring the help of religion and philosophy to understand and all the techniques of magic to manipulate." The public ate it up. Weil lectured to packed audiences and appeared frequently on *Oprah* and *Larry King Live.* His books became international best sellers, and his face appeared on the cover of *Time*—twice. *Publishers Weekly* described Weil as "America's best-known complementary care physician," the *San Francisco Chronicle* as "the guru of alternative medicine," *Time* as "Mr. Natural," and his own books as "America's most trusted medical expert." Andrew Weil is one of America's most famous, most influential alternative healers.

Another of Mehmet Oz's "Superstars" is Deepak Chopra. Chopra was born and raised in New Delhi, where he attended the All India Institute of Medical Sciences, and later moved to the United States to complete residencies in internal medicine and endocrinology. As chief of staff at New England Memorial Hospital, Chopra "noticed a growing lack of fulfillment." He asked himself, "Am I doing all I can for my patients?" So he visited onetime Beatles guru Maharishi Mahesh Yogi, who persuaded Chopra to found the American Association of Ayurvedic Medicine and become the director of the Maharishi Ayurveda Health Center. Ayurvedic medicine, founded in India two thousand years ago, is based on the ancient Greek notion of balancing humors. However, unlike Hippocrates's four humors, ayurvedic medicine balances three humors, or doshas: wind (*vata*), choler (*pitta*), and phlegm (*kapha*). To determine whether doshas are out of balance, healers take a patient's pulse.

Chopra became a national guru on Monday, July 12, 1993, when he appeared on *The Oprah Winfrey Show* to promote his book *Ageless Body, Timeless Mind*. Within twenty-four hours he had sold 137,000 copies; by the end of the week it was 400,000.

In addition to Old Testament and ancient Greek, Chinese, and Indian remedies, Oz also promotes the relatively modern concepts of homeopathy and chiropractic manipulations, both of which represent a kind of devolution in medical thinking.

Homeopathy was the creation of Samuel Hahnemann, who practiced in Germany and France between 1779 and 1843. Hahnemann was disturbed by the brutality of nineteenth-century medicine, which included bloodletting with leeches, poison-induced vomiting, and skin blistering with acids. He wanted a safer, better way to treat people. His epiphany came in 1790. While ingesting powder from the bark of a cinchona tree, Hahnemann developed a fever. At the time, it was known that cinchona bark, which contained quinine, could treat malaria. Hahnemann believed that because he had fever, and because fever was a symptom of malaria, medicines should induce the same symptoms as the disease. For example, vomiting illnesses should be treated with medicines that cause vomiting. (Homeopathy literally means "similar suffering.") To be on the safe side, Hahnemann also believed that homeopathic medicines should be diluted to the point that they aren't there anymore. Although the active ingredient was gone, Hahnemann believed, the final preparation would be influenced by the medicines having once been there.

Like homeopathy, chiropractic manipulations are also the brainchild of one man: Daniel D. Palmer. Palmer was a mesmerist who used magnets to treat his patients. But in 1895, when a man who had been deaf for seventeen years walked into his office, Palmer tried something else. Believing that deafness was caused by a misaligned spinal column, which he called "subluxation," Palmer pushed down on the back of the man's neck, hoping to realign his spine. It worked; the man recovered his hearing immediately. (The event is often referred to as "the crack heard round the world.") Most miraculous about Palmer's cure is that the eighth cranial nerve, which conducts nerve impulses from the ear to the brain, doesn't travel through the neck. Palmer then took the next illogical step, arguing that *all* diseases were caused by misaligned spines. Because this isn't true, it shouldn't be surprising that studies have shown that chiropractic manipulations don't treat many of the diseases they are claimed to, such as headaches, menstrual pain, colic, asthma, and allergies.

Although Oz promotes therapies born before scientists had determined what caused diseases and why, he's enormously popular—for many reasons.

First, Oz and his Superstars provide an instruction book for something that doesn't come with instructions: life. Collectively, books written by Oz, Weil, and Chopra tell people exactly what to eat and when to eat it; how to be a friend; how to sustain a loving relationship; how and when to exercise; which shampoos, cleaning fluids, laundry detergents, and baby foods to use; how to prepare meals (including "Dr. Weil's Favorite Low-Fat Salad Dressing"); and how to treat almost every possible illness. It's reassuring to know that there's a right and wrong way to do everything. And because these books are so definitive, so clear about how to handle almost any disease, they inspire a cultlike devotion among their followers. Do it our way and you'll live longer, love better, and raise happier, healthier children. Given life's arbitrary, capricious, and unpredictable nature, these books can be quite comforting.

Another lure of alternative medicine is that it's personalized. Practitioners of modern medicine can appear callous and insensitive. Patients feel more like a number than a person. That's where alternative healers come in: they provide individual care, because they care. "Doctors are trapped in this system," says Andrew Weil. "A ravenously for-profit system." But Weil isn't trapped: "I listen to them," he says. "I take sixty minutes on a first visit." "My advice for everybody," says Mehmet Oz, "is to customize therapy for yourself."

The promise of ancient wisdom is also appealing. When Mehmet Oz discussed acupuncture on *The Dr. Oz Show*, he made a rather surprising statement. "It's the basis of ancient Chinese medicine," he insisted. Oz was arguing that we should trust ancient medicine *because* it's ancient. Today's culture is filled with this sentiment. For example, in the movie *2012*, starring John Cusack and Amanda Peet, the world is coming to an end—something that apparently had been predicted by the Mayan calendar. "All our scientific advances," laments one scientist, "all our fancy machines—the Mayans saw this coming thousands of years ago." The writers of *2012* knew their audience. Many people believe that ancient healers and soothsayers, free from confusing modern technologies, possessed a clearer, wiser view of things. "One of the arguments mobilized by alternative medicine practitioners against orthodox medicine is that the latter is constantly changing while alternative medicine has remained unaltered for hundreds, even thousands of years," wrote Raymond Tallis in *Hippocratic Oaths: Medicine and Its Discontents*. "The lack of development in 5,000 years can be a good thing only if 5,000 years ago alternative practitioners already knew of entirely satisfactory treatments. If they did, they have been remarkably quiet about them." Modern medicine is carved by centuries of learning. It continues to evolve because it continues to generate new information. It isn't fixed in time. But the fluidity of modern medicine can be unsettling. Alternative medicine's certainty, on the other hand, can be quite reassuring.

Ironically, while alternative remedies are embraced in the developed world, they're often rejected in the countries where they originated. In mainland China, for example, where both traditional and modern therapies are available, only 18 percent of the population relies on alternative medicines; in Hong Kong, 14 percent; and in Japan, even less. In China, acupuncture is embraced almost solely by the rural poor. "It's easy for the well-fed metropolitan with time and money on his hands to talk about dealing with chronic symptoms with ayurvedic medicine or Chinese herbal therapies or ancient African or Native American remedies," writes John Diamond in *Snake Oil and Other Preoccupations*. "But if you go to the countries where those remedies are all they have, you'll find them crying out for good old Western antibiotics, painkillers, and all the rest of the modern and expensive pharmacopoeia. When the government of South Africa complains that not enough is being done to help the 10 percent of its population which is HIV-positive, it isn't asking for help with preparing 'natural' remedies: it wants AZT."

Traditional healers also offer something else. Where modern medicine is spiritless and technological, they argue, alternative medicine is spiritual and meaningful. "The more the universe seems comprehensible," wrote Steven Weinberg, a Nobel Prize–winning physicist, "the more it seems pointless." Although modern science offers the prospect of longer lives, it doesn't offer the prospect of more meaningful lives. Alternative medicine, on the other hand, offers something greater: better health imbued with a deeper sense of purpose. Oz, Weil, and Chopra proffer their remedies with a spirituality that borders on mysticism. "Nothing is more dangerous than science without poetry or technical progress without emotional content," wrote Houston Stewart Chamberlain, a German philosopher. In a culture that doesn't understand technology, and is often frightened and disappointed by it, spiritualism is an easy sell.

Finally, practitioners of alternative medicine appeal to the popular notion that you can manage your own health, that you don't need doctors to tell you what to do. "Alternative medicine is at the grass roots level," says Oz. "And because of that, nobody owns it. Alternative medicine empowers us. And if it does work for you, don't let anybody take it away." The offer of control in a health-care system where patients feel little or no control is irresistible. "The lure of alternative therapies won't end," says Harriet Hall, a former flight surgeon and a regular contributor to *Skeptical Inquirer* magazine, "until you take the 'human' out of human nature."

At the heart of our distrust of modern medicine is the notion that we've rejected nature at our own peril—that big pharmaceutical companies, by synthesizing products in laboratories, have led us away from the natural products that allow us to live longer. And what could be more natural than vitamins.

Part II

THE LURE OF ALL THINGS NATURAL

2
The Vitamin Craze:

Linus Pauling's Ironic Legacy

I gotta tell you, right at the top of my list would be taking vitamins. I know that over the years doctors have said they're ridicu-
lous and all that. But I started taking my vitamins at an early age.
And I take them every day. Every bloody day! So I think that's
number one. For whatever reason, I feel active and pretty good
at my age.

—Regis Philbin

Everyone loves vitamins. Derived from the Latin word *vita*,
meaning "life," vitamins are necessary for the conversion
of food into energy. Millions of Americans believe that taking
daily vitamins makes them feel better and live longer.

Thirteen vitamins have been identified. Nine are easily
dissolved in water: vitamins B_1 (thiamine), B_2 (riboflavin), B_3
(niacin), B_5 (pantothenic acid), B_6 (pyridoxine), B_7 (biotin), B_9

(folic acid), B_{12} (cobalamin), and C (ascorbic acid). Four aren't water-soluble: vitamins A (retinol), D (calciferol), E (tocopherol), and K (phylloquinone). When people don't get enough vitamins, they suffer diseases such as beriberi, pellagra, scurvy, and rickets (caused by deficiencies of vitamins B_1, B_3, C, and D, respectively).

The problem with most vitamins is that they aren't made inside the body; they're available only in foods or supplements. So the question isn't "Do people need vitamins?" They do. The real questions are "How much do they need?" and "Do they get enough in foods?" Nutrition experts and vitamin manufacturers are split on the answers to these questions. Nutrition experts argue that all people need is the recommended daily allowance (RDA), typically found in a routine diet. Industry representatives argue that foods don't contain enough vitamins and that larger quantities are needed. Fortunately, many excellent studies have now resolved the issue.

On October 10, 2011, researchers from the University of Minnesota found that women who took supplemental multivitamins died at rates higher than those who didn't. Two days later, researchers from the Cleveland Clinic found that men who took vitamin E had an increased risk of prostate cancer. "It's been a tough week for vitamins," said Carrie Gann of ABC News.

These findings weren't new. Seven previous studies had already shown that vitamins increased the risk of cancer and heart disease and shortened lives. Still, in 2012, more than half of all Americans took some form of vitamin supplements. What few people realize, however, is that their fascination with vitamins can be traced back to one man. A man who was so spectacularly right that he won two Nobel Prizes and so spectacularly wrong that he was arguably the world's greatest quack.

Linus Pauling was born on February 28, 1901, in Portland, Oregon. He attended Oregon Agricultural College (now Oregon State University), in Corvallis, before entering the California Institute of Technology (Caltech), where he taught for more than forty years.

In 1931, Pauling published a paper in the *Journal of the American Chemical Society* titled "The Nature of the Chemical Bond." Before publication, chemists knew of two types of chemical bonds: ionic, where one atom gives up an electron to another; and covalent, where atoms share electrons. Pauling argued that it wasn't that simple—electron sharing was somewhere between ionic and covalent. Pauling's idea revolutionized the field, marrying quantum physics with chemistry. His concept was so revolutionary that when the journal editor received the manuscript, he couldn't find anyone qualified to review it. When Albert Einstein was asked what he thought of Pauling's work, he shrugged his shoulders. "It was too complicated for me," he said. For this single paper, Pauling received the Langmuir Prize as the most outstanding young chemist in the United States, became the youngest person elected to the National Academy of Sciences, was made a full professor at Caltech, and won the Nobel Prize in Chemistry. He was thirty years old.

In 1949, Pauling published a paper in *Science* titled "Sickle Cell Anemia, a Molecular Disease." At the time, scientists knew that hemoglobin (the protein in blood that transports oxygen) crystallized in the veins of people with sickle-cell anemia, causing joint pain, blood clots, and death. But they didn't know why. Pauling was the first to show that sickle hemoglobin had a slightly different electrical charge—a quality that dramatically affected how the hemoglobin reacted with oxygen. His finding gave birth to the field of molecular biology.

In 1951, Pauling published a paper in the *Proceedings of the National Academy of Sciences* titled "The Structure of Proteins." Scientists knew that proteins were composed of a series of amino acids. Pauling proposed that proteins also had a secondary structure determined by how they folded upon themselves. He called one configuration the alpha helix—later used by James Watson and Francis Crick to explain the structure of DNA.

In 1961, Pauling collected blood from gorillas, chimpanzees, and monkeys at the San Diego Zoo. He wanted to see whether mutations in hemoglobin could be used as a kind of evolutionary clock. Pauling showed that humans had diverged from gorillas about 11 million years ago, much earlier than scientists had suspected. A colleague later remarked, "At one stroke he united the fields of paleontology, evolutionary biology, and molecular biology."

Pauling's accomplishments weren't limited to science. Beginning in the 1950s—and for the next forty years—he was the world's most recognized peace activist. Pauling opposed the internment of Japanese Americans during World War II, declined Robert Oppenheimer's offer to work on the Manhattan Project, stood up to Senator Joseph McCarthy by refusing a loyalty oath, opposed nuclear proliferation, publicly debated nuclear-arms hawks like Edward Teller, forced the government to admit that nuclear explosions could damage human genes, convinced other Nobel Prize winners to oppose the Vietnam War, and wrote the best-selling book *No More War!* Pauling's efforts led to the Nuclear Test Ban Treaty. In 1962, he won the Nobel Peace Prize—the first person ever to win two unshared Nobel Prizes.

In addition to his election to the National Academy of Sciences, two Nobel Prizes, the National Medal of Science, and the Medal for Merit (which was awarded by the president of the United States), Pauling received honorary degrees from Cambridge University, the University of London, and the University of Paris. In 1961, he appeared on the cover of *Time* magazine's Men of the Year issue, hailed as one of the greatest scientists who had ever lived.

Then all the rigor, hard work, and hard thinking that had made Linus Pauling a legend disappeared. In the words of a colleague, his "fall was as great as any classic tragedy."

The turning point came in March 1966, when Pauling was sixty-five years old. He had just received the Carl Neuberg Medal. "During a talk in New York City," recalled Pauling, "I mentioned how much pleasure I took in reading about the discoveries made by scientists in their various investigations of the nature of the world, and stated that I hoped I could live another twenty-five years in order to continue to have this pleasure. On my return to California I received a letter from a biochemist, Irwin Stone, who had been at the talk. He wrote that if I followed his recommendation of taking 3,000 milligrams of vitamin C, I would live not only twenty-five years longer, but probably more." Stone, who referred to himself as Dr. Stone, had spent two years studying chemistry in college. Later, he received an honorary degree from the Los Angeles College of Chiropractic and a "PhD" from Donsbach University, a non-accredited correspondence school in Southern California.

Pauling followed Stone's advice. "I began to feel livelier and healthier," he said. "In particular, the severe colds I had suffered several times a year all my life no longer occurred. After a few years, I increased my intake of vitamin C to ten times, then twenty times, and then three hundred times the RDA: now 18,000 milligrams per day."

From that day forward, people would remember Linus Pauling for one thing: vitamin C.

I n 1970, Pauling published *Vitamin C and the Common Cold*, urging the public to take 3,000 milligrams of vitamin C every day (about fifty times the RDA). Pauling believed that the common cold would soon be a historical footnote. "It will take decades to eradicate the common cold completely," he wrote, "but it can, I believe, be controlled entirely in the United States and some other countries within a few years. I look forward to witnessing this step toward a better world." Pauling's book became an instant best seller. Paperback versions were printed in 1971 and 1973, and an expanded edition titled *Vitamin C, the Common Cold and the Flu*, published three years later, promised to ward off a predicted swine flu pandemic. Sales of vitamin C doubled, tripled, and quadrupled. Drugstores couldn't keep up with demand. By the mid-1970s, 50 million

Americans were following Pauling's advice. Vitamin manufacturers called it "the Linus Pauling effect."

Scientists weren't as enthusiastic. On December 14, 1942, about thirty years before Pauling published his first book, Donald Cowan, Harold Diehl, and Abe Baker, from the University of Minnesota, published a paper in the *Journal of the American Medical Association* titled "Vitamins for the Prevention of Colds." The authors concluded, "Under the conditions of this controlled study, in which 980 colds were treated . . . there is no indication that vitamin C alone, an antihistamine alone, or vitamin C plus an antihistamine have any important effect on the duration or severity of infections of the upper respiratory tract."

Other studies followed. After Pauling's pronouncement, researchers at the University of Maryland gave 3,000 milligrams of vitamin C every day for three weeks to eleven volunteers and a sugar pill (placebo) to ten others. Then they infected volunteers with a common cold virus. All developed cold symptoms of similar duration. At the University of Toronto, researchers administered vitamin C or placebo to 3,500 volunteers. Again, vitamin C didn't prevent colds, even in those receiving as much as 2,000 milligrams a day. In 2002, researchers in the Netherlands administered multivitamins or placebo to more than 600 volunteers. Again, no difference. At least fifteen studies have now shown that vitamin C doesn't treat the common cold. As a consequence, neither the FDA, the American Academy of Pediatrics, the American Medical Association, the American Dietetic Association, the Center for Human Nutrition at the Johns Hopkins Bloomberg School of Public Health, nor the Department of Health and Human

Services recommend supplemental vitamin C for the prevention or treatment of colds.

Although study after study showed that he was wrong, Pauling refused to believe it, continuing to promote vitamin C in speeches, popular articles, and books. When he occasionally appeared before the media with obvious cold symptoms, he said he was suffering from allergies.

Then Linus Pauling upped the ante. He claimed that vitamin C not only prevented colds; it cured cancer.

In 1971, Pauling received a letter from Ewan Cameron, a Scottish surgeon from a tiny hospital outside Glasgow. Cameron wrote that cancer patients who were treated with ten grams of vitamin C every day had fared better than those who weren't. Pauling was ecstatic. He decided to publish Cameron's findings in the *Proceedings of the National Academy of Sciences* (*PNAS*). Pauling assumed that as a member of the academy he could publish a paper in *PNAS* whenever he wanted; only three papers submitted by academy members had been rejected in more than half a century. Pauling's paper was rejected anyway, further tarnishing his reputation among scientists. Later, the paper was published in *Oncology*, a journal for cancer specialists. When researchers evaluated the data, the flaw became obvious: the cancer victims Cameron had treated with vitamin C were healthier at the start of therapy, so their outcomes were better. After that, scientists no longer took Pauling's claims about vitamins seriously.

But Linus Pauling still had clout with the media. In 1971, he declared that vitamin C would cause a 10 percent decrease in deaths from cancer. In 1977, he went even further. "My present estimate is that a decrease of 75 percent can be achieved with vitamin C alone," he wrote, "and a further decrease by use of other nutritional supplements." With cancer in their rearview mirror, Pauling predicted, Americans would live longer, healthier lives. "Life expectancy will be 100 to 110 years," he said, "and in the course of time, the maximum age might be 150 years."

Cancer victims now had reason for hope. Wanting to participate in the Pauling miracle, they urged their doctors to give them massive doses of vitamin C. "For about seven or eight years, we were getting a lot of requests from our families to use high-dose vitamin C," recalls John Maris, chief of oncology and director of the Center for Childhood Cancer Research at the Children's Hospital of Philadelphia. "We struggled with that. They would say, 'Doctor, do you have a Nobel Prize?'"

Blindsided, cancer researchers decided to test Pauling's theory. Charles Moertel, of the Mayo Clinic, evaluated 150 cancer victims: half received ten grams of vitamin C a day and half didn't. The vitamin C–treated group showed no difference in symptoms or mortality. Moertel concluded, "We were unable to show a therapeutic benefit of high-dose vitamin C." Pauling was outraged. He wrote an angry letter to the *New England Journal of Medicine*, which had published the study, claiming that Moertel had missed the point. Of course vitamin C hadn't worked: Moertel had treated patients who had already received chemotherapy. Pauling claimed that vitamin C worked only if cancer victims had received no prior chemotherapy.

Bullied, Moertel performed a second study; the results were

the same. Moertel concluded, "Among patients with measurable disease, none had objective improvement. It can be concluded that high-dose vitamin C therapy is not effective against advanced malignant disease regardless of whether the patient had received any prior chemotherapy." For most doctors, this was the end of it. But not for Linus Pauling. He was simply not to be contradicted. Cameron observed, "I have never seen him so upset. He regards the whole affair as a personal attack on his integrity." Pauling thought Moertel's study was a case of "fraud and deliberate misrepresentation." He consulted lawyers about suing Moertel, but they talked him out of it.

Subsequent studies have consistently shown that vitamin C doesn't treat cancer.

Pauling wasn't finished. Next, he claimed that vitamin C, when taken with massive doses of vitamin A (25,000 international units) and vitamin E (400 to 1,600 IU), as well as selenium (a basic element) and beta-carotene (a precursor to vitamin A), could do more than just prevent colds and treat cancer; they could treat virtually every disease known to man. Pauling claimed that vitamins and supplements could cure heart disease, mental illness, pneumonia, hepatitis, polio, tuberculosis, measles, mumps, chickenpox, meningitis, shingles, fever blisters, cold sores, canker sores, warts, aging, allergies, asthma, arthritis, diabetes, retinal detachment, strokes, ulcers, shock, typhoid fever, tetanus, dysentery, whooping cough, leprosy, hay fever, burns, fractures, wounds, heat prostration, altitude sickness, radiation poisoning, glaucoma, kidney failure, influenza, bladder ailments, stress, rabies, and snakebites. When the

AIDS virus entered the United States in the 1970s, Pauling claimed vitamins could treat that, too.

On April 6, 1992, the cover of *Time*—rimmed with colorful pills and capsules—declared, "The Real Power of Vitamins: New research shows they may help fight cancer, heart disease, and the ravages of aging." The article, written by Anastasia Toufexis, echoed Pauling's ill-founded, disproved notions about the wonders of megavitamins. "More and more scientists are starting to suspect that traditional medical views of vitamins and minerals have been too limited," wrote Toufexis. "Vitamins—often in doses much higher than those usually recommended—may protect against a host of ills ranging from birth defects and cataracts to heart disease and cancer. Even more provocative are glimmerings that vitamins can stave off the normal ravages of aging." Toufexis enthused that the "pharmaceutical giant Hoffman-La Roche is so enamored with beta-carotene that it plans to open a Freeport, Texas, plant next year that will churn out 350 tons of the nutrient annually, or enough to supply a daily 6 milligram capsule to virtually every American adult."

The National Nutritional Foods Association (NNFA), a lobbying group for vitamin manufacturers, couldn't believe its good luck, calling the *Time* article "a watershed event for the industry." As part of an effort to get the FDA off their backs, the NNFA distributed multiple copies of the magazine to every member of Congress. Speaking at an NNFA trade show later in 1992, Toufexis said, "In fifteen years at *Time* I have written many health covers. But I have never seen anything like the response to the vitamin cover. It whipped off the sales racks, and we were inundated with requests for copies. There are no more copies. 'Vitamins' is the number-one-selling issue so far this year."

Although studies had failed to support him, Pauling believed that vitamins and supplements had one property that made them cure-alls, a property that continues to be hawked on everything from ketchup to pomegranate juice and that rivals words like *natural* and *organic* for sales impact: *antioxidant*.

Antioxidation vs. oxidation has been billed as a contest between good and evil. The battle takes place in cellular organelles called mitochondria, where the body converts food to energy, a process that requires oxygen and so is called oxidation. One consequence of oxidation is the generation of electron scavengers called free radicals (evil). Free radicals can damage DNA, cell membranes, and the lining of arteries; not surprisingly, they've been linked to aging, cancer, and heart disease. To neutralize free radicals, the body makes its own antioxidants (good). Antioxidants can also be found in fruits and vegetables—specifically, selenium, beta-carotene, and vitamins A, C, and E. Studies have shown that people who eat more fruits and vegetables have a lower incidence of cancer and heart disease and live longer. The logic is obvious: if fruits and vegetables contain antioxidants—and people who eat lots of fruits and vegetables are healthier—then people who take supplemental antioxidants should also be healthier.

In fact, they're less healthy.

In 1994, the National Cancer Institute, in collaboration with Finland's National Public Health Institute, studied 29,000 Finnish men, all long-term smokers more than fifty years old. This group was chosen because they were at high risk for cancer and heart disease. Subjects were given vitamin E, beta-carotene, both, or neither. The results were clear: those taking vitamins and supplements were *more* likely to die from lung cancer or heart disease than those who didn't take them—the opposite of what researchers had anticipated.

In 1996, investigators from the Fred Hutchinson Cancer Research Center, in Seattle, studied 18,000 people who, because they had been exposed to asbestos, were at increased risk of lung cancer. Again, subjects received vitamin A, beta-carotene, both, or neither. Investigators ended the study abruptly when they realized that those who took vitamins and supplements were dying from cancer and heart disease at rates 28 and 17 percent higher, respectively, than those who didn't.

In 2004, researchers from the University of Copenhagen reviewed fourteen randomized trials involving more than 170,000 people who took vitamins A, C, E, and beta-carotene to see whether antioxidants could prevent intestinal cancers. Again, antioxidants didn't live up to the hype. The authors concluded, "We could not find evidence that antioxidant supplements can prevent gastrointestinal cancers; on the contrary, *they seem to increase overall mortality.*" When these same researchers evaluated the seven best studies, they found that death rates were 6 percent higher in those taking vitamins.

In 2005, researchers from Johns Hopkins School of Medicine evaluated nineteen studies involving more than 136,000 people and found an increased risk of death associated with supplemental vitamin E. Dr. Benjamin Caballero, director of the Center for Human Nutrition at the Johns Hopkins Bloomberg School of Public Health, said, "This reaffirms what others have said. The evidence for supplementing with any vitamin, particularly vitamin E, is just not there. This idea that people have that [vitamins] will not hurt them may not be that simple." That same year, a study published in the *Journal of the American Medical Association* evaluated more than 9,000 people who took high-dose vitamin E to prevent cancer; those who took vitamin E were *more* likely to develop heart failure than those who didn't.

In 2007, researchers from the National Cancer Institute examined 11,000 men who did or didn't take multivitamins. Those who took multivitamins were twice as likely to die from advanced prostate cancer.

In 2008, a review of all existing studies involving more than 230,000 people who did or did not receive supplemental antioxidants found that vitamins increased the risk of cancer and heart disease.

On October 10, 2011, researchers from the University of Minnesota evaluated 39,000 older women and found that those who took supplemental multivitamins, magnesium, zinc, copper, and iron died at rates higher than those who didn't. They concluded, "Based on existing evidence, we see little justification for the general and widespread use of dietary supplements."

Two days later, on October 12, researchers from the Cleveland Clinic published the results of a study of 36,000 men who took vitamin E, selenium, both, or neither. They found that those receiving vitamin E had a 17 percent greater risk of prostate cancer. In response to the study, Steven Nissen, chairman of cardiology at the Cleveland Clinic, said, "The concept of multivitamins was sold to Americans by an eager nutraceutical industry to generate profits. There was never any scientific data supporting their usage." On October 25, a headline in the *Wall Street Journal* asked, "Is This the End of Popping Vitamins?"

Studies haven't hurt sales. In 2010, the vitamin industry grossed $28 billion, up 4.4 percent from the year before. "The thing to do with [these reports] is just ride them out," said Joseph Fortunato, chief executive of General Nutrition Centers. "We see no impact on our business."

How could this be? Given that free radicals clearly damage cells—and given that people who eat diets rich in substances that neutralize free radicals are healthier—why did studies of supplemental antioxidants show they were harmful? The most likely explanation is that free radicals aren't as evil as advertised. Although it's clear that free radicals can damage DNA and disrupt cell membranes, that's not always a bad thing. People need free radicals to kill bacteria and eliminate new cancer cells. But when people take large doses of antioxidants, the balance between free radical production and destruction might tip too much in one direction, causing an unnatural state in which the immune system is less able to kill harmful invaders. Researchers have called this "the antioxidant paradox." Whatever the reason, the data are clear: high doses of vitamins and supplements increase the risk of heart disease and cancer; for this reason, not a single national or international organization responsible for the public's health recommends them.

In May 1980, during an interview at Oregon State University, Linus Pauling was asked, "Does vitamin C have any side effects on long-term use of, let's say, gram quantities?" Pauling's answer was quick and decisive. "No," he replied. About one year later, Pauling's wife, Ava Helen, who had been taking massive doses of vitamins, died from stomach cancer.

Despite a wealth of scientific evidence, most Americans don't know that megavitamins are unsafe. So why don't more people know about this? And why hasn't the FDA sounded an alarm? The answer is predictable: money and politics.

Part III

LITTLE SUPPLEMENT MAKERS VERSUS BIG PHARMA

3

The Supplement Industry Gets a Free Pass:

Neutering the FDA

Liberty for the wolves is death for the lambs.
—Isaiah Berlin

Government oversight of the pharmaceutical industry has been a long, tortuous journey filled with unimaginable tragedies. "The story of drug regulation," wrote historian Michael Harris, "is built on tombstones."

It started with purveyors of patent medicines.

"How much is your health worth, ladies and gentlemen? It's priceless, isn't it? Well, my friends, one half-dollar is all it takes to put you in the pink. That's right, ladies and gents. For fifty pennies, Nature's True Remedy will succeed where doctors have failed. Only Nature can heal and I have Nature right here in this little bottle. My secret formula, from God's own laboratory, the Earth itself, will cure rheumatism, cancer, diabetes, baldness, bad breath, and curvature of the spine."

In the 1800s, medical hucksters could claim anything. Boston Drug cured drunkenness. Pond's Extract treated meningitis. Hydrozone prevented yellow fever. Peruna calmed inflammation of the ovaries. Liquozone cured asthma, bronchitis, cancer, dysentery, eczema, gallstones, hay fever, malaria, and tuberculosis. And Dr. Williams' Pink Pills for Pale People treated all that and more. Sales were limited only by what customers were willing to believe. By the turn of the century, patent medicines were a $75-million-a-year business. It didn't last. On June 30, 1906, the federal government stepped in, passing the Pure Food and Drug Act. Three men led the charge; one was concerned about foods, another about drugs, the third about neither.

H arvey Washington Wiley was raised on a farm in southern Indiana. He studied classics at nearby Hanover College before serving as a corporal in the Civil War. Although Wiley later graduated from Indiana Medical College, he never practiced medicine. Rather, he pursued his love of science, obtaining a degree in chemistry from Harvard. In 1874, Wiley became the chair of the chemistry department at the newly opened Purdue University. Ten years later, he was chief chemist at the United States Department of Agriculture (USDA), at a time when the food industry, like the drug industry, was unregulated. Wiley watched helplessly as Americans consumed spoiled meat, sawdust-adulterated flour, and formaldehyde-preserved milk. It was time, Wiley argued, for the federal government to step in.

S amuel Hopkins Adams graduated from Hamilton College before joining the staff of the *New York Sun*, one of the nation's most influential newspapers. On October 7, 1905, Adams published the first of a series of articles in *Collier's* magazine titled "The Great American Fraud." Adams wanted Americans to know what they were buying. So he sent samples of patent medicines to chemists, finding that many contained large quantities of alcohol: Paine's Celery Compound contained 21 percent; Peruna, 28 percent; and Hostetter's Stomach Bitters, 44 percent. (To put this in perspective, beer contains 4 to 6 percent alcohol, wine 10 to 15 percent, and whiskey 35 to 45 percent.) Patent medicine makers were in the liquor business. They were also in the narcotics business. Adams found that several medicines contained opium, morphine, hashish, and cocaine. These drugs were often given to babies; Winslow's Soothing Syrup, for example, was loaded with morphine. When Adams asked his maid how she had left her small children alone at night, she replied, "They're all right. Just one teaspoon of Winslow's and they lay like dead until morning." Perhaps the best example of the subterranean narcotics industry was Coca-Cola, introduced in 1886 as an "Intellectual Beverage and Temperance Drink" that offered the virtues of cocaine without the stigma of alcohol.

By the last installment of "The Great American Fraud," in

February 1906, Samuel Adams had exposed 264 companies and individuals, listed scores of people who had died from dangerous drugs, and shown that many patent medicines had caused diseases rather than treated them. "Every man who trades in this market, whether he pockets the profits of the maker, the purveyor, or the advertiser, takes [his] toll of blood," wrote Adams. "Here the patent medicine business is its naked-est, most cold-hearted. Relentless greed sets the trap, and death is partner in the enterprise." More than 500,000 Americans read "The Great American Fraud."

With the public up in arms about Adams's publication, Harvey Wiley felt the time was right. He proposed a federal law to "cover every kind of medicine for external and internal use," which would require manufacturers to list all ingredients and prohibit them from selling narcotics without a prescription. Wiley's proposal angered the Proprietary Association of America, lobbyists for the industry. "Such a law," advised its Committee on Legislation, "would practically destroy the sale of proprietary remedies in the United States." Industry executives successfully lobbied to kill the bill.

And that would have been the end of it had it not been for a die-hard socialist who, if anything, wanted less government, not more.

Upton Sinclair was an unknown journalist who railed against the sins of American capitalism. In the early 1900s, he traveled to Chicago to write a fictional work about the plight of immigrant workers in the meatpacking industry. With *The Jungle*, Sinclair wanted to inspire his readers; instead he nauseated them. "There would be meat that had tumbled on the floor, in the dirt and sawdust, where the workers had tramped and spit uncounted billions of consumption germs," wrote Sinclair. "There would be meat stored in great piles in rooms; and the water from leaky roofs would drip over it, and thousands of rats would race about on it. It was too dark in these storage places to see well, but a man could run his hand over these piles of meat and sweep off handfuls of the dried dung of rats. These rats were nuisances, and the packers would put poisoned bread out for them; they would die and then rats, bread and meat would go into the hoppers together." Sinclair described how employees occasionally slipped into steaming vats, later emerging as Durham's Pure Leaf Lard. Wanting to hit Americans in their hearts, he hit them in their stomachs. Sales of meat dropped by half. Following publication of *The Jungle*, Theodore Roosevelt ordered Congress to create legislation guaranteeing clean meat and pure food.

T he bill that President Roosevelt signed into law, the Pure Food and Drug Act of 1906, was a watered-down version of what Harvey Wiley had wanted. If a patent medicine contained alcohol, cocaine, opium, chloroform, or other potentially harmful drugs, manufacturers had to print it on the label. They could still sell narcotics and dangerous drugs; they just had to tell consumers they were doing it. Most important, no statement could be made that was "false or misleading." Although the law didn't ask manufacturers to prove that their medicines were safe or effective, it was a start. The federal government now had a hand in regulating the drug industry.

Enforcement of the Pure Food and Drug Act fell to the USDA's Bureau of Chemistry. In 1927, the newly minted Food, Drug, and Insecticide Administration took over; three years later, it changed its name to the Food and Drug Administration.

The next federal law was born of the worst pharmaceutical disaster in United States history. It involved one of the first antibiotics: sulfanilamide. In the early 1930s, six companies made sulfa drugs: Squibb, Merck, Winthrop, Eli Lilly, Parke-Davis, and the S. E. Massengill Company of Bristol, Tennessee. Massengill made it poorly. To make sulfa more palatable for children, Harold Watkins, Massengill's chief chemist, suspended it in diethylene glycol. The final preparation—called Elixir Sulfanilamide—contained diethylene glycol, sulfanilamide, water, and small amounts of raspberry extract, saccharin, caramel, and amaranth, which gave the drug a deep reddish purple color. Unlike other sulfa preparations, Massengill's tasted great—perfect for children. The drug, however, was far from perfect, and Massengill knew it. Ten months before marketing the mixture, chemists at Massengill found that a 3 percent solution of diethylene glycol caused fatal kidney failure in rats; Elixir Sulfanilamide contained 72 percent.

In September 1937, Massengill distributed 240 gallons of its elixir in the United States. Three hundred fifty people drank it and immediately suffered heartburn, nausea, cramps, dizziness, vomiting, diarrhea, and difficulty breathing. Even worse, more than a hundred people died from kidney failure, thirty-four of them young children. Following the tragedy, the president of Massengill said, "My chemists and I deeply regret the fatal results, but there was no error in the manufacture of the product. We have been supplying legitimate professional demand and not once could have foreseen the unlooked-for results. I do not feel there was any responsibility on our part." Massengill's president, knowing that his company had acted within the law, wasn't particularly remorseful. Harold Watkins, the chemist who had formulated the product, was. Soon after the incident he killed himself.

The Elixir Sulfanilamide disaster led to the next major drug law: the Food, Drug, and Cosmetic Act of 1938. Now the FDA required safety testing *before* drugs were sold. The newer, stronger law stated that drugs, cosmetics, and therapeutic devices had to be proved safe; manufacturing plants had to be registered and inspected by the FDA every two years; and foods sold across state lines had to be pure and wholesome, safe to eat, and produced under sanitary conditions. Violators could be imprisoned for a year, with longer sentences for second offenses or fraud.

Although the Food, Drug, and Cosmetic Act of 1938 tightened the reins, manufacturers still didn't have to prove that their products worked before selling them. It took another tragedy to make that happen.

On October 1, 1957, Chemie Grünenthal, a West German pharmaceutical company, distributed a sedative called thalidomide. Advertisements claimed that it was safe, even for pregnant women. Within three years, hundreds of women in Europe had delivered babies whose hands and feet were attached directly to their bodies—a disorder called phocomelia (cruelly referred to by the press as "flipper babies"). As many as 24,000 fetuses were damaged by thalidomide; half died before birth. Although Chemie Grünenthal had submitted its drug for licensure in the United States, Dr. Frances Kelsey, an FDA physician, turned it down, believing that the first few reports of phocomelia following the introduction of thalidomide weren't a coincidence.

Because of Kelsey, thalidomide was never marketed in the United States. Still, the disaster led to the next important federal drug law: the 1961 Kefauver-Harris Amendment to the Food, Drug, and Cosmetic Act. The amendment included several new regulations: manufacturers now had to show that drugs were not only safe but *effective* before licensure (even though the thalidomide disaster had nothing to do with the drug's effectiveness); previously licensed drugs could be withdrawn if they were found to be unsafe; manufacturers had to obtain consent from patients before testing experimental drugs; prescription drug advertisements had to include a summary of possible side effects; product labels had to list exact quantities of *all* ingredients; and manufacturers had to adhere to the code of Good Manufacturing Practices in testing, processing, packaging, and storing drugs. For the first time in history, it appeared that America's Magical Miracle Medicine Show would be closing its tent.

Then drug regulation took a giant step backward.

I n 1970, Linus Pauling published his book advising people to take 3,000 milligrams of vitamin C every day, about fifty times the recommended amount. Although the FDA didn't mind people taking vitamins, Pauling's advice scared them; no one really knew whether massive doses were safe. In December 1972, the FDA announced its plan to regulate vitamins containing more than 150 percent of the recommended amount; those containing larger quantities would require proof of safety before sale. Vitamin makers saw this as a threat to their $700 million-a-year business. Represented by the National Health Federation (NHF), the industry set out to destroy the bill. In the end, it did far more than that.

The NHF was (and is) enormously influential. Founded in 1955 and headquartered in Monrovia, California, it consisted of vitamin industry executives and their lobbyists. The list of NHF founders, officers, and board members reads like a who's who of American quackery:

- Harry Hoxsey, who helped found the NHF, made his fortune selling arsenic, pepsin, potassium iodide, and laxatives to treat cancer before fleeing to Tijuana to escape a fraud conviction.
- Fred Hart, president of the Electronic Medical Foundation and the NHF's co-founder, distributed electronic devices that he claimed treated hundreds of diseases before a United States district court ordered him to stop.
- Royal Lee, who owned and operated the Vitamin Products Company, served on the board of governors. Lee published a book claiming that polio could be prevented

by diet alone, even though a polio vaccine had already been invented. One FDA official said Royal Lee was "probably the largest publisher of unreliable and false nutritional information in the world."

- Kirkpatrick Dilling, the NHF's lawyer, was also a lawyer for the Cancer Control Society, a group that promoted questionable cancer cures (which in part explains his interest in representing John and Mary Hofbauer in their attempts to treat their son with laetrile).

- Bruce Halstead, another NHF leader, was convicted of twenty-four counts of fraud for claiming that an herbal tea called ADS treated cancer. ADS, a brownish sludge containing water and bacteria typically found in human feces, sold for $125 to $150 a quart. A Los Angeles County deputy district attorney called Halstead "a crook selling swamp water." After his license to practice medicine was revoked, Halstead was fined $10,000 and sentenced to four years in prison.

- Victor Earl Irons, vice chairman of NHF's board of governors, made Vit-Ra-Tox, a vitamin mixture sold door-to-door. Irons's company created the Vit-Ra-Tox 7-Day Cleansing Program, which included fasting, supplements, herbal laxatives, and a daily enema of strong black coffee. "If every person in this country took two to three home colonics a week," said Irons, "95 percent of the doctors would have to retire for lack of business." Irons received a one-year prison sentence for making false claims.

In addition to lobbying for the unrestricted sale of megavitamins, the NHF also campaigned against pasteurization, vaccination, and fluoridation.

Although Linus Pauling was a key to blocking the FDA's attempt to regulate megavitamins, industry executives knew they needed a political insider to win the day—someone who would not only defeat the bill requiring safety studies of megavitamins but free them from FDA regulation entirely. It didn't take long for the NHF to find its man: Senator William Proxmire, a Democrat from Wisconsin. Proxmire was best known for his Golden Fleece Awards, given to federally funded science programs he considered wasteful. One winner, the Aspen Movie Map project, spawned a technology that enabled soldiers to familiarize themselves quickly with new territory. Although Proxmire later apologized to several award winners, his name became a verb: "to Proxmire" meant to obstruct scientific research for political gain. In 1975, William Proxmire introduced a bill banning the FDA from regulating megavitamins. Bob Dole, William Fulbright, Barry Goldwater, Hubert Humphrey, George McGovern, and Sam Nunn were cosponsors.

On the morning of August 14, 1974, Senator Edward Kennedy, chair of the Senate Committee on Health, Education, Labor and Pensions, called the meeting to order. "The Food and Drug Administration, in my opinion, has an obvious and important responsibility to protect the American consumer against foods and drugs [that] are potentially harmful," he said. "It must make certain that Americans are not led to believe that dietary products are therapeutic or in some way beneficial, when in fact they may be worthless and a waste of money." Proxmire was the first to defend his bill, claiming that the recommended daily allowance for vitamins was far too low: "What the FDA wants to do is to strike the views of its stable of orthodox nutritionists into tablets and bring them down from Mount Sinai where they will be used to regulate the rights of millions of Americans. The real issue is whether the FDA is going to play God."

Others rose in support of Proxmire's bill. Bob Dole, who would later appear in television ads for Viagra, said, "I would like to be on record as absolutely opposing any action by the FDA to regulate the retail sale of vitamin and mineral nutrients. In fact, it's a little inconceivable to me that such restrictions should ever have been promulgated in the first place." Milton Bass, a lawyer proficient in the doublespeak of his industry, said, "The Proxmire bill is designed for one purpose. It is designed to permit the customer to buy a safe food product honestly labeled." Bass failed to explain how defeating legislation requiring proof of safety made products safer.

Representing the FDA was its commissioner, Dr. Alexander Schmidt. Kennedy asked Schmidt to respond to Proxmire's contention that vitamins weren't harmful at any dosage. "Well, the word *harm* is relative," said Schmidt. "What is overlooked by a great many people is that while there is not a lot of evidence that very large doses of water-soluble vitamins are harmful, there is not a lot of information that large doses of water-soluble vitamins are safe either." Absence of evidence, argued Schmidt, wasn't evidence of absence.

Schmidt wasn't alone in his opposition. Dr. Sidney Wolfe, representing Ralph Nader's consumer advocacy group, Public

Citizen, said, "This is a drug industry. The difference between large doses of vitamins and over-the-counter [drugs] is non-existent. Exploitation of genuine concerns people have for their health [by promoting] vitamin pill–popping solutions is no better than . . . fraud." Marsha Cohen, an attorney with Consumers Union, made a plea for common sense. Setting eight cantaloupes in front of her, she said, "We can safely rely upon the limited capacity of the human stomach to protect persons from overindulgence in any particular vitamin- or mineral-rich food. For example, you would have to eat eight cantaloupes to take in barely 1,000 milligrams of vitamin C. But just these two little pills, easy to swallow, contain the same amount. . . . And 1,000 milligrams, it should be recalled, is on the low end of Dr. Pauling's recommended 250 to 10,000 milligrams daily. If the proponents of the legislation before you succeed, one tablet would contain as much vitamin C as all of these cantaloupes—or even twice, thrice or twenty times that amount. And there would be no protective satiety level." Cohen had pointed to the vitamin industry's Achilles' heel: ingesting large quantities of vitamins was unnatural, the opposite of what manufacturers had been promoting.

Kennedy, Schmidt, Wolfe, and Cohen were supported by the American Association of Retired Persons (AARP), the American Academy of Pediatrics, and the American Society of Clinical Nutrition. It didn't matter. On September 24, 1974, Proxmire's bill passed by a vote of 81 to 10. On April 23, 1976, it became law. "It was the most humiliating defeat in the history of the FDA," wrote Peter Barton Hutt, the FDA's chief counsel. Dan Hurley, author of *Natural Causes: Death, Lies, and Politics in America's Vitamin and Herbal Supplement Industry*, wrote, "Congress had made the decision to roll back the government's authority over the sale of foods and drugs for the first time in the twentieth century. So began an unprecedented experiment to test whether the unbridled use of vitamins and other supplements would help or hinder health, with the American public as the guinea pig."

B y the early 1990s, the Proxmire Amendment had opened a dangerous door—one that extended well beyond the uncontrolled sale of megavitamins. Henry Waxman, a Democratic congressman from California, and David Kessler, the FDA's newly appointed commissioner, wanted to close it.

Waxman and Kessler were concerned that salespeople in health food stores were advising customers to treat high blood pressure, infections, and cancer with vitamins, supplements, minerals, and herbs. "Unsubstantiated claims are becoming more exaggerated," said Kessler. "We are back at the turn of the century, when snake oil salesmen could hawk their potions with promises that couldn't be kept. If you walk into a health food store, you have to recognize that we have not approved the safety of these products nor substantiated their claims."

On June 7, 1991, Henry Waxman introduced the Food, Drug, Cosmetic, and Device Enforcement Amendments, which "authorized any district court to order the recall of a food, drug, device, or cosmetic which is in violation of the [law] if the violation involves fraud or presents a significant risk to human or animal health." The word most important to Waxman and Kessler—and that most frightened the supplement industry—was *fraud*. The FDA knew that claims of safety and effectiveness by supplement manufacturers were either unsubstantiated or wrong; it wanted to protect the consumer by making the industry prove it. Otherwise, the American public would continue to be hoodwinked. But Waxman and Kessler had leaned into a left hook. By taking on a wealthy, powerful, politically connected industry, they not only didn't get what they'd wanted; they got the opposite of what they'd wanted. "Kessler wanted to drive a stake into the heart of the dietary supplement industry," recalled Peter Barton Hutt. "Instead, he drove it into the heart of the FDA."

R ising to meet Waxman's bill was Gerry Kessler (no relation to David), the founder of one of the nation's most successful supplement companies, Nature's Plus. On February 22, 1991, Gerry Kessler asked seventy industry leaders to attend a meeting at his home, near Santa Barbara. Executives had to be impressed by the opulence around them. Called the Circle K Ranch, Kessler's home boasted a 17,000-square-foot main lodge, a koi pond, tennis courts, swimming pools, visitors' cabins, a fitness center, and flocks of wandering ostriches and trumpeter swans. The estate's previous occupant had been Ray Kroc, the owner of McDonald's.

Industry insiders sent their wealthiest, most influential representatives to Kessler's meeting. Present were Allen Skolnick, of Solgar Vitamin and Herb Company, which later sold for half a billion dollars; Milton Bass, Scott Bass, and Martie Whittekin, representing the National Nutritional Foods Association, the health food industry's lobbying group; Sandy Gooch, founder of Mrs. Gooch's, the leading natural-food seller in the western United States; and Scott Randolph, from the conglomerate that owned Nature's Bounty, Rexall Sundown, Puritan's Pride, and Vitamin World—a group with sales of more than $1 billion a year.

Gerry Kessler was an aggressive, brilliant, persuasive man. His goal was to convince each company to put up hundreds of thousands of dollars to turn the argument around. Kessler must have known that he couldn't defeat the FDA by proving his products' claims. His best chance was to persuade the American public that what the FDA really wanted was to limit their freedom. Let's appeal to the public's desire to get the government off their backs, he argued, to buy what they want when they want it. "We've got to deal with this now," warned Kessler. "We've got to organize. We have to convince the grass roots, the consumers, to get on the bandwagon and fight." It was a brilliant marketing strategy. Soon Gerry Kessler would convince millions of Americans that it was in their best interest not to know what they were buying.

Although science wasn't on his side, Kessler knew that politics could still win the day. And he knew just the politician to do it. In fact, he didn't have to look any further than the end of the table during his meeting with industry insiders. Sitting among the CEOs, lobbyists, and lawyers were Patricia Knight and Jack Martin, top aides to Orrin Hatch, the Republican senator from Utah. Knight would soon become Hatch's chief of staff, Martin an industry lobbyist.

Orrin Hatch loved the supplement industry. As a young man, he sold vitamins and supplements; as an older one, he took them every day—including saw palmetto, to shrink his prostate. "I really believe in them," he said. "I use them daily. They make me feel better, as they make millions of Americans feel better. And I hope they give me that little added edge as we work around here."

In turn, the supplement industry loved Orrin Hatch. Four of the industry's top thirty manufacturers—Weider, Nutraceutical Corporation, Nature's Way, and Nu Skin International—were located in Utah. (It was the only state with its own supplement trade association: the Utah Natural Products Alliance.) At the time of Gerry Kessler's crusade, Utah benefited from several billion dollars in profits from supplement sales. Hatch's campaigns also benefited. As Dan Hurley describes in *Natural Causes*, between 1989 and 1994 Herbalife International gave Hatch $49,250; MetaboLife, $31,500; and Rexall Sundown, Nu Skin International, and Starlight International a total of $88,550. In addition, according to his financial disclosures for 2003, Hatch owned 35,621 shares of Pharmics, a Utah-based nutritional supplement company. In the early 1990s, Hatch's son Scott began working for lobbying groups representing vitamin and supplement makers. Kevin McGuiness, Hatch's former chief of staff, was also a lobbyist for the industry.

With millions of dollars in hand, Gerry Kessler ran a campaign in the 1990s that—like the tobacco campaigns of the 1950s—was a model for how industries could cash in at the expense of the public's health. First, Kessler convinced supplement manufacturers to send preprinted letters to Congress urging freedom of choice in health products. Second, he turned ten thousand health food stores into political-action centers, convincing employees to offer discounts to their customers if they sent letters to Congress. Third, he recruited celebrities. "Start screaming at Congress and the White House not to let the FDA take our vitamins away," said actress Sissy Spacek. Other actors, like Mariel Hemingway, Victoria Principal, and James Coburn, appeared at industry trade shows. Fourth, he made it personal, busing supporters to shout down Henry Waxman at public meetings. "They dominated the meetings," recalled Waxman. "One person after another would harangue me." One of Waxman's office windows was pelted with tomatoes.

Later, Waxman summed up Gerry Kessler's campaign against his amendment. "It was unlike any other lobbying campaign I've ever seen," he said. "People believed what they were being told because it fed into their view that doctors and pharmaceutical companies wanted to block alternative medicines that could keep people healthy. What they didn't understand, though, was that this view was manipulated by people who stood to make a lot of money, and did make a lot of money—billions of dollars."

In the heat of battle, Gerry Kessler met David Kessler in his office. "Commissioner," said Gerry, "I don't know whether you're for us or against us, but I will be here when you're gone, and so will this industry. So it would be good if you were for us." "Is that a threat?" replied David. "No," said Gerry. "I just hope you'll be for us." Gerry Kessler was right. It wasn't a threat. It was a fact. And what was about to happen would mirror the events of the mid-1970s, when William Proxmire turned an attempt to regulate megavitamins into an amendment that allowed vitamin makers to make even more outrageous claims. Now, with the money, power, and greed of a multibillion-dollar-a-year industry behind him, Gerry Kessler would accomplish the same thing with supplements, minerals, and herbs, allowing manufacturers to avoid FDA oversight and effectively hide critical information. It was called the Dietary Supplement Health and Education Act, a name that fit comfortably into the upside-down world of alternative medicine, where one thing often means another—in this case, an education act that had nothing to do with education. Quite the opposite. Now consumers would have no way of knowing whether what they were buying was safe or effective. It is remarkable, even in retrospect, that consumers not only chose not to know what they were buying; they lobbied for it.

Although effective, Gerry Kessler's campaign was amateurish, relying on letter writing and bullying. He needed a lobbyist who had clout, especially with the Democrats. The recommendation that turned the tide came from Bill Richardson, a Democrat from New Mexico, a state with a long-standing interest in alternative medicine. Richardson suggested that Gerry Kessler call Tony Podesta, one of the most influential Democratic lobbyists in Washington and the brother of John Podesta, who would later become Bill Clinton's chief of staff. Although Edward Kennedy had opposed the Supplement Act, he was indebted to Tony Podesta, who had done much to further Kennedy's political agenda. Podesta convinced Kennedy to provide a hearing for Gerry Kessler's bill.

On July 29, 1993, the House hearing for the Dietary Supplement Health and Education Act (DSHEA) was called to order. FDA commissioner David Kessler was one of the first to testify. Standing next to a chart listing serious side effects caused by supplements, he said, "Think about it. Half our prescription drugs are derived from plants, and no one doubts for a minute that drugs can have toxic effects. That is why we insist on rigorous testing to separate out those with unacceptable toxicity. We must not assume that all risk disappears when plants are sold as dietary supplements for therapeutic purposes."

Also testifying against the bill was Dorothy Wilson, a victim of the L-tryptophan disaster. In 1989, L-tryptophan, an amino acid sold in health food stores, had caused a neurological condition that affected more than five thousand people and killed twenty-eight. "Mr. Chairman," said Wilson, who was wheelchair-bound, "before you weaken the statute and make it more difficult for the FDA to do its job, consider how my life was catastrophically and permanently changed. The Dietary Supplement Health and Education Act must not pass. Reject this act because it is the right, courageous, and honorable thing to do." The Consumers Union, the AARP, the American Cancer Society, the American Heart Association, the American Nurses Association, and the American College of Physicians also opposed the bill.

I n August, while Hatch's bill was being considered, the campaign against FDA regulation reached a new low, running a nationwide television ad featuring Mel Gibson. It began with white letters on a black background: "Los Angeles, 9:57 p.m." Police dressed in black clothing, wearing night-vision goggles and carrying rifles burst through a door and run into a bathroom. Mel Gibson's bathroom. "Hey, guys—guys!" shouts Gibson. "Its only vitamins." A warning appears: "The federal government is actually considering classifying most vitamins and other supplements as drugs. The FDA has already conducted raids on doctors' offices and health food stores. Could raids on individuals be next?" As the police handcuff Gibson, he says, "Vitamin C. You know, like in oranges?" Then the take-home message: "Protect your rights to use vitamins and other supplements. Call Congress now." At the end of Gibson's commercial, other celebrities join in. Eddie Albert, star of the television sitcom *Green Acres*, says, "That can actually happen. Don't let them take your supplements away." Whoopi Goldberg says it's "the right of American citizens to have free access to dietary supplements of their choice." Talk show host Jenny Jones says, "We need to ensure that Congress continues to give consumers the right to make intelligent choices about our health."

Perhaps the hearing's most dramatic moment occurred when David Kessler and Orrin Hatch went toe-to-toe—a face-off between a de facto representative of the supplement industry and a doctor who was trying to protect the public. The exchange centered on the FDA's recent seizure of primrose oil due to bogus claims.

> HATCH: What safety hazard was the FDA addressing that warranted such intensive use of agency resources and personnel?
>
> KESSLER: Senator, I can read you the claims made for oil of evening primrose. The list starts with cancer.
>
> HATCH: Remember, the issue is safety I am talking about.

KESSLER: My real concern is the types of diseases for which oil of evening primrose is promoted.

HATCH: But my question is: What proof do you have that this substance is unsafe?

KESSLER: This is being promoted for a lot of different diseases, anywhere from hypertension to atopic dermatitis.

HATCH: Safety, Doctor, safety! This is the question! Is an American citizen more likely to die from an adverse reaction to a drug approved by the FDA or a dietary supplement?

KESSLER: Senator, I am amazed. What do you think are *in* pharmaceuticals? Half our pharmaceuticals come from plants. There are chemicals in pharmaceuticals and those chemicals are found naturally.

Kessler was making an argument that had been made for centuries. The source of a chemical doesn't matter; only the chemical matters. And whether it is synthesized by a pharmaceutical company or found in nature, the chemical is the same. And it should be regulated in the same way. Otherwise consumers will think they're getting a guarantee of safety when they're not.

In the end, industry money trumped common sense. On May 11, 1994, the Dietary Supplement Health and Education Act became law. The act defined a supplement as "a product intended to supplement the diet that bears or contains one or more of the following ingredients: a vitamin, a mineral, an herb or other botanical, or an amino acid." "Breathtaking in its dimensions," wrote Dan Hurley, "[the act] would end forever the simple legal dichotomy between 'food' and 'drug' to create a third, hermaphroditic category that was both yet neither: the dietary supplement. And beyond the usual suspects—vitamins, minerals, herbs, and amino acids—the law would permit manufacturers to define a product as a 'dietary supplement' merely by saying so, no matter how artificially derived. Put lamb's brain in a drug or food, and prepare to spend millions of dollars and a few years on studies showing that it is safe and effective; put it in a supplement and you're good to go, no evidence necessary." The *New York Times* called it the "Snake Oil Protection Act."

O ne way to judge the effect of the 1975 Proxmire Amendment and the 1994 Dietary Supplement Health and Education Act is to compare two products: vitamins and Vioxx.

On May 20, 1999, the FDA approved Merck's Vioxx for the treatment of pain and arthritis. The drug was an immediate success, grossing $2.5 billion a year.

On November 23, 2000, the *New England Journal of Medicine* published a study comparing Vioxx with the anti-inflammatory drug Naprosyn for the treatment of pain. Merck, which funded the study, wanted to prove that its drug was less likely than Naprosyn to cause intestinal bleeding. It was. But there was a side effect that hadn't been anticipated: patients receiving Vioxx were four times as likely to have a heart attack. The authors of the study argued that because Naprosyn inhibited the aggregation of platelets (a factor in the blockage of the arteries supplying blood to the heart) and Vioxx didn't, Vioxx wasn't causing heart attacks; it just wasn't preventing them. The FDA didn't buy it, sending a warning letter to Merck's CEO, Ray

Gilmartin. The FDA was angry that Merck hadn't adequately warned consumers about the possibility that Vioxx might cause heart attacks. Merck complied. By April 2002, every vial of Vioxx contained a warning label.

Merck had hoped that when Vioxx was compared with something that didn't have an anti-platelet effect, it wouldn't increase the risk of heart attacks. On March 17, 2005, another study was published in the *New England Journal of Medicine*, again supported by Merck; this time Vioxx was compared to a placebo. Investigators found that the risk of heart attacks was lower in the second study (twofold rather than fourfold); and it was a problem only after people had used the drug for at least eighteen months. But the risk was still there. So on September 23, 2004, months before the results of the second study were published, Merck voluntarily withdrew Vioxx from the market. "We are taking this action because we believe it best serves the interests of patients," said Gilmartin. "Although we believe it would have been possible to continue to market Vioxx with labeling that would incorporate these new data, given the availability of [other] therapies and the questions raised by the data, we concluded that a voluntary withdrawal is the responsible course to take."

Because the FDA was watching, consumers were well aware of problems caused by Vioxx.

So which is more dangerous: Vioxx or vitamins? Indeed, both have dangers. The better question is, why does everybody know that Vioxx can cause heart disease and nobody knows that megavitamins can cause cancer? The answer is that we have chosen not to know.

Americans are constantly manipulated by public relations campaigns. Sometimes these campaigns champion messages that make our lives better (stop smoking); sometimes they make our lives worse (start smoking). But you have to take your hat off to a campaign that has us not only asking for something that hurts us but screaming for it. Such were the campaigns by the National Health Federation for the Proxmire Amendment and by the National Nutritional Foods Association for the DSHEA. In both cases, the industry manipulated the media, politicians, celebrities, and the public to stoke a multibillion-dollar-a-year industry. And in both cases, the industry forced the FDA to back off so it could make claims about safety and efficacy that weren't true. When Vioxx was found to cause heart attacks, the FDA issued press releases that were immediately picked up by the media. But the FDA doesn't regulate vitamins and supplements, so it can't warn the public when there's a problem. Although the supplement industry skillfully manipulated people into believing that the Supplement Act was about freedom to make health choices, it was really about freeing the supplement industry to offer unsafe choices. If knowledge is power, DSHEA is powerlessness.

The vitamin and supplement industry had successfully created a false dichotomy. On one side are natural products: vitamins, minerals, dietary supplements, plants, and herbs. Because they're natural, they're safe. On the other side are drugs. Because drugs are man-made, they're supposedly more dangerous. However, many drugs are derived from nature, including antibiotics. Furthermore, the notion that natural products aren't dangerous is fanciful. Fava beans (*Vicia faba*) can cause severe anemia; castor beans contain ricin, the most potent neurotoxin known to man; jimsonweed contains hallucinogenic alkaloids; berries from the coyotillo plant (*Karwinskia humboldtiana*) cause paralysis; and the ackee fruit (*Blighia sapida*) causes a severe lowering of blood sugar ("Jamaican vomiting sickness"). Mother Nature can kill you. "Just because something is natural it does not mean that it is good," write Simon Singh and Edzard Ernst in *Trick or Treatment*, "and just because something is unnatural it does not mean that it is bad. Arsenic, cobra poison, nuclear radiation, earthquakes, and the ebola virus can all be found in nature, whereas vaccines, spectacles, and artificial hips are all man-made."

The possibility of harm caused by natural products sold in health food stores isn't theoretical. Blue cohosh can cause heart failure; nutmeg can cause hallucinations; comfrey, kava, chaparral, *Crotalaria*, *Senecio*, jin bu huan, *Usnea* lichen, and valerian can cause hepatitis; monkshood and plantain can cause heart arrhythmias; wormwood can cause seizures; stevia leaves can decrease fertility; concentrated green tea extracts can damage the liver; milkweed seed oil and bitter orange (*Citrus aurantium*) can cause heart damage; thujone can cause neurological damage; and concentrated garlic can cause bleeding. Indeed, one of the worst dietary supplement disasters in history occurred in 1992, when about a hundred people developed kidney failure from a "slimming" mixture found to contain the plant aristolochia; at least seventy patients required kidney transplants or dialysis, and many later developed bladder cancers. In 2008, more than two hundred people—including a four-year-old—were poisoned by massive doses of selenium contained in Total Body Formula and Total Body Mega. The products were supposed to contain 200 micrograms of selenium per serving; instead they contained 40,800 micrograms. Herbal remedies can also cause harm: two infants died from a tea containing pennyroyal and another from a decongestant containing capsaicin. In October 2013, the Hawaii Department of Health began an investigation of a supplement called OxyElite Pro that was linked to twenty-four cases of hepatitis; eleven people were hospitalized, two received liver transplants, and one died. Because the dietary supplement industry is essentially unregulated, only 170 (0.3 percent) of the 54,000 products on the market have documented safety tests.

And it's not just the supplements themselves that might be harmful, but what's contaminating them. In 2004, researchers at Harvard Medical School tested Indian (Ayurvedic) remedies obtained from shops near Boston's City Hall. They found that 20 percent contained potentially harmful levels of lead, mercury, and arsenic. Between 1978 and 2004, herbal medicines caused fifty-five cases of severe or fatal heavy-metal poisoning. In late 2009, Kirkman Labs, a supplement manufacturer popular among parents of children with autism, recalled fifteen thousand bottles of zinc because they contained undeclared antimony, a heavy metal.

In 2012, the FDA estimated that approximately 50,000 adverse reactions to supplements occurred every year. In 2013, the supplement industry experienced several disasters. In August, vitamin and mineral preparations made by Purity First Health Products in Farmingdale, Connecticut, were found to contain dimethazine and methasterone: two powerful anabolic steroids. The problem came to light when twenty-nine people were sickened; women developed masculinizing symptoms, such as hair loss, lowering of voices, and loss of menstrual periods, and men reported impotence and low testosterone. In September, it was found that a dietary supplement called OxyElite Pro had caused twenty-four cases of acute hepatitis; two people required liver transplants and one died. In November, researchers in Ontario discovered that popular herbal products like echinacea, St. John's wort, and ginko biloba often contained other herbs instead, some of which could be quite dangerous. By the end of the year, at least forty-two dietary supplements had been found to be contaminated with prescription drugs.

Some argue, reasonably, that drugs made by pharmaceutical companies and licensed by the FDA can also cause serious or fatal side effects. For example, antibiotics can cause severe allergic reactions, and chemotherapeutic drugs can suppress the immune system, leading to devastating and occasionally fatal infections. But the side effects from licensed products are known and contained in package inserts. Doctors prescribing these drugs weigh the risks and benefits and monitor the patient for side effects. Because dietary supplements aren't tested for safety before they are put on the market, safety problems are evident only after the products have been sold and patients have been harmed. Also, while it would surely be safer to treat cancer with homeopathy, acupuncture, and supplements instead of chemotherapy, these therapies wouldn't work. Serious diseases require serious medicines.

Worst of all: most Americans don't realize what's happened. A Harris poll found that 68 percent believe that the government requires herbal manufacturers to report side effects, 58 percent believe that the FDA must approve herbal products before sale, and 55 percent believe that manufacturers of vitamins, minerals, and dietary supplements cannot make claims about safety or effectiveness without scientific evidence. All these beliefs couldn't be further from the truth. Serious problems caused by dietary supplements are the industry's dirty little secret. Because of the Supplement Act, they'll remain a secret.

n 2007, as problems with the industry continued to mount, FDA regulators were finally granted permission to supervise the way supplements were made. Although they still couldn't force manufacturers to prove that their products were safe or effective, at least they could make sure the product contained what the label said it contained. What the FDA found was appalling. Of the 450 supplement manufacturers inspected, at least half had significant problems. One, ATF Fitness, substituted ingredients without changing the product label. Others didn't even have recipes for their products. And some manufactured products in buildings contaminated with rodent feces and urine—in one facility a rodent was found cut in half next to a scoop. "It's downright scary," said Daniel Fabricant, head of the FDA's Division of Dietary Supplement Programs. "At least half of the industry is failing on its face." Cara Welch, a vice president for the National Products Association, an industry trade group, called the findings "unfortunate."

n 1994, when the Supplement Act passed, sales of dietary supplements were $4 billion a year; by 2007, $28 billion; by 2012, $34 billion. America's Magical Miracle Medicine Show is back in business.

4

Fifty-Four Thousand Supplements:

Which Ones Work?

There cannot be two kinds of medicine. There is only medicine that has been adequately tested and medicine that has not.

—Marcia Angell, former editor-in-chief of the *New England Journal of Medicine*

Health food stores are wonderlands of promise. If people want to burn fat, detoxify livers, shrink prostates, avoid colds, stimulate brains, boost energy, reduce stress, enhance immunity, prevent cancer, extend lives, enliven sex, or eliminate pain, all they have to do is walk in. The question, however, is which products work? And how do we know they work?

Fortunately, thanks to James Lind, we can figure it out. When

Lind climbed aboard the HMS *Salisbury*—intent on finding a cure for scurvy—he moved medicine from a faith-based system to an evidence-based system. No longer do we have to *believe* in treatments. Now we can test them to see whether they work. For example, alternative healers recommend ginkgo or rose and orange oils for memory; graviola, astragalus, and cat's claw for immunity; guarana and *Cordyceps* for energy; chicory root for constipation; lemon balm oil, ashwagandha, eleuthero, Siberian ginseng, and holy basil for stress; sage and black cohosh for menstrual pain; coconut oil and curry powder for Alzheimer's disease; saw palmetto for prostate health; sandalwood bark to prevent aging; garlic for high cholesterol; peppermint oil for allergies; artichoke extract and green papaya for digestion; echinacea for colds; chondroitin sulfate and glucosamine for joint pain; milk thistle for hepatitis; St. John's wort for depression; and tongkat ali for sexual potency. Although the size and cost of clinical studies have increased dramatically since the days of James Lind, the claims made by alternative healers are testable—eminently testable.

When pharmaceutical companies make drugs and biologicals, the rules are clear. Company scientists first test the product in animals. If the results are promising, they take the next step, testing it in progressively larger numbers of people. If the results are still promising, they perform a definitive (so-called Phase III) study proving that the product is safe and that it works. For example, two rotavirus vaccines are distributed in the United States (I'm the co-inventor of one of them). These vaccines prevent a common cause of diarrhea and dehydration in infants. Before the existence of a rotavirus vaccine, about 70,000 children in the United States were hospitalized with dehydration caused by rotavirus every year; in the developing world, rotavirus killed 2,000 children every day.

Making a rotavirus vaccine wasn't easy. The Phase III trial for one rotavirus vaccine, RotaTeq, included more than 70,000 children from eleven countries, tested for four years at a cost of about $350 million. If stacked one on top of another, patients' records from that trial would have exceeded the height of the Sears Tower in Chicago. The FDA allowed the manufacturer of RotaTeq to make claims about safety and effectiveness only after those claims had been supported by rigorous scientific studies; otherwise, it wouldn't have licensed the product.

The situation for plants, herbs, and dietary supplements is different. Because of the Supplement Act, the FDA doesn't regulate them, so they don't have to be tested before they're sold. Sometimes supplements are tested by the National Center for Complementary and Alternative Medicine (NCCAM), a branch of the National Institutes of Health. One difference between the FDA and NCCAM is that the FDA requires products to be tested *before* they're sold, whereas NCCAM might test some products after they've been put on the market. If researchers funded by NCCAM find that dietary supplements don't work or have harmful side effects, they publish their results in scientific journals. No product recall. No change in the label. No FDA warnings. If people don't read scientific journals, they won't know that claims on the label are false and misleading.

The driving force behind the creation of NCCAM was Tom Harkin, a popular senator from Iowa who believed his allergies had been cured by eating bee pollen. Harkin figured that the only reason alternative remedies hadn't been brought into the mainstream was that they hadn't been properly tested. Once they were tested and everyone could see that they really worked, alternative medicine would be embraced by modern science and paid for by insurance companies. Since its birth, in 1999, NCCAM officials have spent about $1.6 billion studying alternative therapies. They've spent $374,000 of taxpayer money to find out that inhaling lemon and lavender scents doesn't promote wound healing; $390,000 to find out that ancient Indian remedies don't control Type 2 diabetes; $446,000 to find that magnetic mattresses don't treat arthritis; $283,000 to discover that magnets don't treat migraine headaches; $406,000 to determine that coffee enemas don't cure pancreatic cancer; and $1.8 million to find out that prayer doesn't cure AIDS or brain tumors or improve healing after breast reconstruction surgery. Fortunately, NCCAM has recently abandoned these kinds of studies, choosing instead to focus on studies of dietary supplements and pain relief.

Let's assume for a moment that everyone wants what's best for the patient. Alternative healers believe that ancient medicine is of value. And mainstream doctors and pharmaceutical companies believe that modern science has the most to offer. Peter Medawar, a Nobel Prize–winning immunologist, calls the battle for recognition by those who promote various remedies a "kindly conspiracy." "Exaggerated claims for the efficacy of a [therapy] are very seldom the consequence of any intention to deceive," he writes. "They are usually the outcome of a kindly conspiracy in which everybody has the best intentions. The patient wants to get well, his physician wants to have made him better, and the pharmaceutical company would like to put it into the physician's power to have made him so. *The controlled clinical trial is an attempt to avoid being taken in by this conspiracy of goodwill.*"

Using Medawar's logic, terms like *conventional* and *alternative* medicine are misleading. If a clinical trial shows that a therapy works, it's not an alternative. And if it doesn't work, it's also not an alternative. In a sense, there's no such thing as alternative medicine. For example, Hippocrates used the leaves of the willow plant to treat headaches and muscle pains. By the early 1800s, scientists had isolated the active ingredient: aspirin. In the 1600s, a Spanish physician found that the bark of the cinchona tree treated malaria. Later, cinchona bark was shown to contain quinine, a medicine now proven to kill the parasite that causes malaria. In the late 1700s, William Withering used the foxglove plant to treat people with heart failure. Later, foxglove was found to contain digitalis, a drug that increases heart contractility. More recently, artemisia, an herb used by Chinese healers for more than a thousand years, was found to contain another anti-malaria drug, which was later called artemisinin. "Herbal remedies are not really alternative," writes Steven Novella, a Yale neurologist. "They have been part of scientific medicine for decades, if not centuries. Herbs are drugs and they can be studied as drugs. My problem is with the regulation and marketing of specific herbal products, because they often make claims that are not backed by evidence." Unfortunately, when natural products promoted by alternative healers have been put to the test, they've often fallen far short of their claims.

Although mainstream medicine hasn't found a way to treat dementia or enhance memory, practitioners of alternative medicine claim that they have: ginkgo biloba. As a consequence, ginkgo is one of the ten most commonly used natural products, netting hundreds of millions of dollars a year for its manufacturers. Unfortunately, sales exceed claims. Between 2000 and 2008, the National Institutes of Health funded a collaborative study by the University of Washington, the University of Pittsburgh, Wake Forest University, Johns Hopkins University, and the University of California at Davis to determine whether ginkgo worked. More than 3,000 elderly adults were randomly assigned to receive ginkgo or a placebo (a sugar pill). Decline in memory and onset of dementia were the same in both groups. In 2012, a study of more than 2,800 adults found that ginkgo didn't ward off Alzheimer's disease.

Another example is St. John's wort. Every year, ten million people suffer major depression in the United States, and every year 35,000 people kill themselves. For each successful suicide, eleven more have tried. Depression is a serious illness; to treat it, scientists have developed medicines that alter brain chemicals such as serotonin. Called selective serotonin reuptake inhibitors (SSRIs), these drugs are licensed by the FDA. Because they've been shown to help with severe depression, doctors recommend them. Practitioners of alternative medicine, however, have a better idea—a more natural, safer way to treat severe depression: St. John's wort. Because so many people use St. John's wort, and because severe depression, if not properly treated, can lead to suicide, NCCAM studied it. Between November 1998 and January 2000, eleven academic medical centers randomly assigned 200 outpatients to receive St. John's wort or a placebo, finding no difference in any measure of depression.

Another favorite home remedy is garlic to lower cholesterol. Because high cholesterol is associated with heart disease, because heart disease is a leading cause of death, because lipid-lowering agents lower cholesterol, and because many people are choosing garlic instead of lipid-lowering agents, researchers studied it. In 2007, Christopher Gardner and coworkers at Stanford University School of Medicine evaluated the effects of garlic on 192 adults with high levels of low-density lipoprotein cholesterol (bad cholesterol). Six days a week for six months, participants received either raw garlic, powdered garlic, aged garlic extract, or a placebo. After checking cholesterol levels monthly, investigators concluded, "None of the forms of garlic used in this study . . . had statistically or clinically significant effects on low-density lipoprotein cholesterol or other plasma lipid concentrations in adults with moderate hypercholesterolemia." In other words, patients who choose garlic to treat their bad cholesterol are choosing to do nothing for a problem that could lead to severe and even fatal heart disease.

S aw palmetto is also popular. As a man ages, his prostate enlarges, which blocks the flow of urine. If untreated, prostate enlargement can cause urinary tract infections, bladder stones, and kidney failure. Fortunately, medicines that relax muscles within the prostate or reduce its size have been available for years. But practitioners of alternative medicine prefer saw palmetto; more than 2 million men use it.

In 2006, NCCAM supported a study at the University of California at San Francisco, the San Francisco Veterans Affairs Medical Center, and Northern California Kaiser Permanente. Investigators assigned 225 men with moderate to severe symptoms of prostate enlargement to receive either saw palmetto or a placebo twice daily for a year, finding no difference between the two groups in urinary flow rate, prostate size, or quality of life.

Five years later, the study was repeated, this time with higher doses. Researchers from Washington University School of Medicine, in St. Louis, studied 369 men who received increasing doses of saw palmetto or a placebo. Again, no change in urinary symptoms. "Now we know that even very high doses of saw palmetto make absolutely no difference," said study author Gerald Andriole, chief of urologic surgery at the school. "Men should not spend their money on this herbal supplement as a way to reduce symptoms of an enlarged prostate because it clearly does not work any better than a sugar pill."

A choice to believe the hype about saw palmetto was a choice to risk the occasionally severe complications of prostate enlargement. Again, natural wasn't better. It was worse.

Another popular remedy is milk thistle. "A most interesting tonic herb from the tradition of European folk medicine is milk thistle, *Silybum marianum*," wrote Andrew Weil in 1995. "The seeds from this plant yield an extract, silymarin, that enhances metabolism of liver cells and protects them from toxic injury. I recommend this herb to all patients with chronic hepatitis and abnormal liver function." Unfortunately, Weil's recommendation didn't stand up to scientific study. In 2011, Dr. Michael Fried, of the University of North Carolina at Chapel Hill, led a group of investigators in determining whether milk thistle helped patients with chronic hepatitis C. More than 150 people infected with hepatitis C virus were given either milk thistle or a placebo. Then investigators determined the amount of liver damage, as well as the quantities of hepatitis C virus in blood. They found no difference between the two groups.

Alternative healers also recommend chondroitin sulfate and glucosamine for joint pain. In 2006, Daniel Clegg, of the University of Utah, led a group of investigators to see whether it worked. They studied more than 1,500 people who were given either chondroitin sulfate alone, glucosamine alone, both, a placebo, or Celebrex (an FDA-licensed anti-inflammatory drug). Only Celebrex worked.

Perhaps the most popular herbal remedy in the United States is echinacea. Used to treat colds, it's a $130-million-a-year business. In 2003, James Taylor and coworkers at the University of Washington, in Seattle, studied more than 400 children with colds who had received either echinacea or a placebo for ten days. The only difference: children taking echinacea were more likely to develop a rash.

Not all the news is grim. Some supplements are clearly of value. For example, all newborns should receive a vitamin K shot at birth. If they don't, about 1 percent of children will develop vitamin K deficiency bleeding, which can be quite severe. Strict vegetarians should supplement their diet with about 2.5 micrograms of vitamin B12 every day. Babies who are exclusively or partially breast-fed should receive 400 IU a day of supplemental vitamin D because it isn't contained in human milk and because they don't go out in the sun much.

Five other supplements have also been promoted for otherwise healthy people: omega-3 fatty acids to prevent heart disease; calcium and vitamin D for postmenopausal women to prevent bone thinning; folic acid during pregnancy to prevent birth defects; and multivitamins for people who are concerned that they aren't getting what they need in their diet.

L ike vitamins, omega-3 fatty acids aren't made in the body, so they have to come from other sources. Some studies have shown that omega-3s protect against high blood pressure and heart disease; others haven't. The best place to get them is in the diet, specifically in fatty fish such as salmon, in vegetable oils such as soybean, rapeseed (canola), and flaxseed, and in walnuts. To get enough omega-3 fatty acids, the American Heart Association recommends that people eat at least one serving of fatty fish at least twice a week. Most Americans consume about 1.6 grams of omega-3s every day, well above what is needed to maintain heart health.

In 2013, two published studies cast doubt on the value and safety of supplemental omega-3 fatty acids. In one excellent, well-controlled study of more than 12,500 people published in the *New England Journal of Medicine*, supplemental omega-3 fatty acids were not found to lessen the incidence of heart disease in those at highest risk. Another study, published in the *Journal of the National Cancer Institute*, found that high levels of omega-3 fatty acids in the bloodstream correlated with the development of prostate cancer. Taken together, these studies significantly dampened enthusiasm for the use of supplemental omega-3 fatty acids.

C alcium is the most abundant mineral in the body, required for vascular tone, muscle function, nerve transmission, and hormone secretion. Of interest, less than 1 percent of the total body calcium is necessary for performing these functions. The remaining 99 percent is stored in bones, where it supports bone structure and function. The problem with calcium occurs when people get older.

In children and teenagers, bone formation exceeds bone breakdown. In early and middle adulthood, these two processes occur at equal rates. Past the age of fifty, however, especially in postmenopausal women, bone destruction exceeds bone formation. This problem isn't trivial. When bones get thinner (a condition called osteoporosis), they break more easily. About one in three postmenopausal women will fracture their spines, and one in five will fracture their hips. Indeed, every year more than 1.5 million fractures occur in the United States because of bone thinning. The best way to avoid this problem is to eat calcium-containing dairy products such as milk, yogurt, and cheese. Calcium can also be found in calcium-fortified fruit juices, beverages, tofu, and cereals.

To lessen the risk of bone thinning, postmenopausal women are advised to eat diets rich in calcium. Because most women get enough calcium in their diet, and because supplementary calcium has not been shown to reduce fractures in otherwise healthy postmenopausal women, the United States Preventive Services Task Force does not recommend supplemental calcium.

Vitamin D and calcium are linked. People who take in adequate amounts of calcium might still have a problem with bone strength if they do not also get sufficient amounts of vitamin D. That is because vitamin D helps the body absorb calcium from the gut. The good news is that vitamin D is readily made in the skin when exposed to sunlight. To get an adequate amount of vitamin D, people need only expose their face, arms, hands, or back to sunlight (without sunblock) for ten to fifteen minutes a day at least twice a week. This will provide the 600 IU of vitamin D recommended by the Institute of Medicine.

Some people, however, either don't get out into the sun much or live in climates where there isn't much sunlight. For this reason, many foods are supplemented with vitamin D, such as milk, bread, pastries, oil spreads, breakfast cereals, and some brands of orange juice, yogurt, margarine, and soy beverages. Because most people get enough vitamin D in their foods or from exposure to sunlight, the United States Preventive Services Task Force does not recommend supplemental vitamin D. One study, however, published in the *New England Journal of Medicine* in 2012, found that elderly adults receiving 800 IU of vitamin Daily had a lower incidence of bone fractures.

F olic acid is a B-complex vitamin necessary for the production of red blood cells. Without folic acid, people develop anemia. But that's not the biggest problem. Researchers have shown that folic acid deficiency can cause something far worse: severe birth defects. Pregnant women deficient in folic acid have delivered babies with malformations of the spine, skull, and brain. To avoid folic acid deficiency, people need about 400 micrograms a day.

Foods rich in folic acid include vegetables such as spinach, broccoli, lettuce, turnip greens, okra, and asparagus;

fruits such as bananas, melons, and lemons; and beans, yeast, mushrooms, beef liver and kidney, orange juice, and tomato juice. Although there are plenty of sources of this nutrient, many pregnant women weren't getting enough folic acid in their diets. So on January 1, 1998, the FDA required manufacturers to add folic acid to breads, breakfast cereals, flours, cornmeals, pastas, white rice, bakery items, cookies, crackers, and some grains. Now it's almost impossible to become folic-acid deficient. Nonetheless, women are advised to take 400 micrograms of folic acid every day, obtained either from foods or supplements or both. Because about half of pregnancies are unplanned and because birth defects occur very early in pregnancy, all women of childbearing age should make sure they're getting enough folic acid.

Finally, what about multivitamins? Although it is clear that large quantities of vitamins in excess of the recommended daily allowance can do harm, multivitamins don't usually contain excess amounts; they contain approximately the recommended amount. Most people take multivitamins because they are concerned that they aren't getting what they need in their diets. The multivitamin serves as an insurance policy. In November 2013, a review of twenty-six studies found that otherwise healthy men and women didn't have a lesser risk of cancer or heart diseases after taking a daily multivitamin. One heart specialist from the United States Preventative Services Task Force summed it up best: "A healthy, balanced diet is critical for good health," she said, "and that's probably the most important way that we get the nutrients that are essential." In other words, most people wrongly assume that vitamins taken in concentrated pill form are processed and presented to the body in the same way that they would be if you got them in fruits and vegetables. Probably the smartest (and least expensive) way to get what we need is just by paying a little more attention to our diet. Harvard Medical School has a wonderful website that describes how to get all the vitamins and minerals you need from food based on various calorie intakes.

In the end, if a medicine works (like folic acid to prevent birth defects), it's valuable, and if it doesn't work (like saw palmetto to shrink prostates), it's not. "There's a name for alternative medicines that work," says Joe Schwarcz, professor of chemistry and the director of the Office for Science and Society at McGill University. "It's called medicine."

Part IV

WHEN THE STARS SHINE ON ALTERNATIVE MEDICINE

Menopause and Aging:

Suzanne Somers Weighs In

Do not go gentle into that good night.
Rage, rage against the dying of the light.
—Dylan Thomas

Celebrities sell products. Queen Latifah sells CoverGirl makeup. Snoop Dogg sells Pepsi Max. Shaquille O'Neal sells Comcast cable. And Snooki sells Wonderful Pistachios. And—because we trust celebrities—we buy what they're selling. Expertise has nothing to do with it. We don't buy products because we think Queen Latifah is a beauty expert, or Snoop Dogg a beverage expert, or Shaq a media expert, or Snooki a nutrition expert. We buy because we enjoy their acting, singing, dunking, and whatever it is that Snooki does.

Celebrities also offer medical advice. Larry King tells us that ginkgo improves memory; Tom Cruise that psychiatry is a pseudoscience; Roger Moore that duck liver may cause Alzheimer's disease; and British model Heather Mills that meat stays in our colon for forty years. Indeed, celebrities as diverse as Pamela Anderson, Cindy Crawford, Jude Law, David Beckham, Paul McCartney, Prince Charles, and Cher have trumpeted the benefits of homeopathic remedies for decades.

B ut when it comes to selling alternative medical products, perhaps no celebrity has been more financially successful than Suzanne Somers.

Somers first appeared as the blonde in the Thunderbird in *American Graffiti* and as a pool girl in Clint Eastwood's *Magnum Force*. In 1978, she got her first big break, landing the role of Chrissy Snow on ABC's *Three's Company*, co-starring Joyce DeWitt and John Ritter. At the beginning of the fifth season, Somers alienated her co-stars. Incensed that male actors like Alan Alda and Carroll O'Connor were making more money, she demanded a pay raise from $30,000 to $150,000 per episode. When the producers refused, Somers boycotted, claiming a broken rib. At the end of the season, she was fired and replaced by Jenilee Harrison. Somers sued ABC for $2 million, unsuccessfully, but she quickly found other work. From the mid-1980s into the 1990s, she starred in the sit-coms *She's the Sheriff* and *Step by Step*. She also performed in Las Vegas. But her most recognized role was as the spokesperson for Thigh-Master, "the best way to tone, shape, and firm your inner thighs with just a few squeezes a day."

In 2001, after undergoing conventional treatment for breast cancer (lumpectomy and radiation therapy), Somers chose an alternative medicine for her post-cancer care. "I chose a nonconventional way to treat my cancer," she told Larry King. "At the time they said, 'Okay, here's what we're going to do . . . Tamoxifen, which is the after-care drug, which, oh, by the way, is going to make you gain weight. And oh, by the way, you'll probably get a little depressed for five years.' And I thought, *It just doesn't sound like a great option.* And then I found a medicine that builds up your immune system. And I thought, *Build up or poison?* So, I decided to go this other way." Instead of Tamoxifen Somers chose Iscador, an alternative remedy made from mistletoe.

Breast cancer had brought Suzanne Somers into the world of alternative medicine. But it was menopause that made her a crusader.

W hen it hit me, it was like a Mack truck," Somers told Oprah eight years after her battle with breast cancer. "It was on my fiftieth birthday. And that began a three-year odyssey of not sleeping, of moodiness, of weight gain, of changes in my hair, changes in my skin." On the same show, Oprah asked Christiane Northrup, a gynecologist, to explain what had happened. "When someone refers to hot flashes, mood swings, irritability, irregular periods, they're really referring to the perimenopausal transition. It is usually a five-to-eight-year process, and it begins with the start of irregular periods, usually by age forty-five." In simple terms, the ovaries stop producing two hormones: estrogen and progesterone. Somers's description was more entertaining: "Suddenly the Seven Dwarfs of Menopause arrived at my door without warning: Itchy, Bitchy, Sweaty, Sleepy, Bloated, Forgetful, and All-Dried-Up."

Somers was frustrated by the medical establishment's inability to provide relief. "As I went from doctor to doctor," she wrote, "I realized I was on my own. No doctor seemed to understand this passage [into menopause]. . . . I wanted a natural, effective way to deal with this. . . . [T]hen I found the solution—a cutting-edge endocrinologist/anti-aging doctor who prescribed a treatment." The effect was immediate. "Wow!" wrote Somers. "What a difference in my life. I was sleeping again. I was happy again. I stopped itching, and bitching, and crying, and, best of all, getting fat." Somers was a convert. She wanted "to scream it from the rooftop." She wrote books about it. She told Larry King about it on CNN. She told Rosie O'Donnell, who later talked about it on *The Joy Behar Show*.

> BEHAR: Are you still in menopause? You used to have hot flashes every day when we were on *The View*.
>
> O'DONNELL: Done. Started at 41, ended at 44. After I was off *The View* Suzanne Somers called me up and said, "You know, I have a feeling that some of your rage [can be cured]."
>
> BEHAR: Really.
>
> O'DONNELL: I'm like, "Are you kidding me?" She's like, "No, really, why don't you try this." And I went to one of her doctors, and I have been putting that cream on me for the last four years and I feel a thousand percent better.
>
> BEHAR: And the rage is gone?
>
> O'DONNELL: Kind of.

Then Somers told Oprah about it. "After one day . . . I felt the veil lift," Oprah wrote in her magazine. "After three days, the sky was bluer, my brain was no longer fuzzy; my memory was sharper. I was literally singing and had a skip in my step." For Suzanne Somers, Rosie O'Donnell, and Oprah Winfrey, the symptoms of menopause had been cured. What was this wondrous medicine?

Treatment of menopause has undergone several shifts. Initially, the approach seemed obvious: replace estrogen and progesterone. "In the 1960s, a book by Robert Wilson called *Feminine Forever* became very popular," Christiane Northrup told Oprah. "It touted the benefits of estrogens as the panacea. Estrogens became the magic bullet that everyone should go on."

As it turns out, it wasn't that easy. Although replacement hormones worked, they came with a price. In 2002, investigators from the Women's Health Initiative—part of the National Institutes of Health—studied the effects of estrogen and progesterone in 17,000 women. Researchers had initially planned to follow women for eight years, but the study was cut short when they noticed a dramatic increase in breast cancer. And it wasn't only breast cancer; replacement hormones also increased the risk of heart disease, strokes, and blood clots. As a consequence, hormone replacement therapy became something doctors began to fear, not embrace.

Women were at a loss. But Suzanne Somers had an answer—an answer that gave birth to a billion-dollar industry. "What was it that sent those wretched dwarfs packing?" wrote Somers. "Natural bioidentical hormones." Somers believed that replacement hormones caused heart disease, blood clots, and cancer because they were made by big pharmaceutical companies—they weren't natural. If women used hormones found in plants—and made by little compounding pharmacies—they could rid themselves of the Seven Dwarfs of Menopause without risk.

Supported by Oprah Winfrey, promoted by Suzanne Somers, and backed by gynecologists like Christiane Northrup, bioidentical hormones have become a national phenomenon. There are, however, a few flaws in the logic.

First, estrogen is estrogen. Whether it's isolated from soybeans, wild yams, or horse's urine, it's the same molecule; the source is irrelevant. The only thing that matters is the molecular structure of the final product. "The implication is that [bioidentical hormones are] something better and something different," said Lauren Streicher, an assistant professor of obstetrics and gynecology at Northwestern University's Feinberg School of Medicine, in Chicago. "Chemically, the structure is exactly the same [as] the FDA products." "Now, repeat after me," wrote McGill's Joe Schwarcz. " 'The properties of a substance depend on molecular structure, not ancestry. When it comes to assessing effectiveness and safety, whether the substance is synthetic or natural is totally irrelevant.' "

Second, the distinction between Big Pharma and small compounding pharmacies, while appealing to the public, is misleading. "They [bioidentical and conventional hormones] are primarily all made at the same factory in Germany," says Streicher. "There's a couple [of large factories] in the United

States. They're the ones that synthesize it from plants and then send it to [small] compounding pharmacies *and* to the major pharmaceutical companies."

If bioidentical and conventional hormones are the same products made in the same place, then they probably carry the same risks. "The big marketing approach of the bioidentical industry is that you can have your cake and eat it," says Wulf Utian, a professor of obstetrics and gynecology at Case Western University. "That these products are not like those made by pharmaceutical companies. They have all the benefits, but they carry none of the risks. And if you believe in that, you believe in the Tooth Fairy."

The difference between bioidentical and conventional hormones isn't that one is natural and the other isn't. Or that one is safe and the other isn't. It's that one is the product of an unsupervised industry and the other isn't. "There's this sense that they're not dangerous," says Streicher, "that people don't have to be monitored. I think they should be regulated so we know the quality that you're getting." Streicher has reason for concern. In 2001, the FDA analyzed twenty-nine products from twelve compounding pharmacies and found that 34 percent failed standard quality or potency tests. For these reasons, agencies responsible for the public's health haven't embraced the bioidentical hormone revolution. The American College of Obstetricians and Gynecologists, the American Association of Clinical Endocrinologists, the American Medical Association, the American Cancer Society, the Mayo Clinic, and the FDA have all issued statements asserting that bioidentical hormones are probably as risky as their conventional counterparts.

Suzanne Somers's discovery that bioidentical hormones could treat menopause was only the beginning. Soon she realized they could offer more. Much more. "We age because our hormones decline," she wrote. "Our hormones don't decline because we age." Bioidentical hormones, argued Somers, could turn back the clock. "You don't want to be sick, do you? You don't want to get fat, shrink, lose your energy, your sex drive, and your brain, or contract any of the diseases that seem to be part and parcel of aging, right? You don't want to end up with bones too feeble to hold up your body. You don't want to walk around with an oxygen tank attached firmly to your back. You don't want to [be] put out to pasture by your family much the same way they do old horses because you are in the beginning of advanced stages of Alzheimer's, do you? But guess what? The second half of your life can be better than the first half. A better life, a healthier life, a life of youthful energy comes from embracing this new medicine. And bioidentical hormone replacement is a big component." But it was far from the only component.

Somers's anti-aging regimen isn't easy. "When I wake up, I start with estrogen every day of the month," she explained to Oprah. "Two weeks of every month, I take progesterone. This is my estrogen arm. This is my progesterone arm." Then she moves to estriol (a form of estrogen). "The other thing that I inject is estriol—two milligrams of this every day vaginally, and I'm not showing you how I do that." Then Somers swallows pills containing calcium, magnesium, folic acid, coenzyme Q10, glucosamine, vitamin C, Eskimo fish oil, omega-3 fatty acids, Flora Source, Adrenal-180 ("because my adrenals were blown out"), SAMe, St. John's wort, L-tryptophan, primrose oil, L-glutamine, carnitine, L-tyrosine, L-taurine, lecithin, glycine, phosphatidylserine, Smoke Shield, rhodiola, white tea capsules, Host Defense, Zyflamend, holy basil, Turmeric Force, selenium, zinc, LycoPom, reishi mushroom, cinnamon, LuraLean, *Phaseolus vulgaris, Irvingia,* green tea phytosome, curcumin, gamma-linoleic acid, resveratrol, vitamin E, vitamin D, and vitamin K_2. She also injects herself with human growth hormone and B-complex vitamins. Then she rubs "a little glutathione cream on the skin on top of my liver to stimulate it." Finally, just to be sure, Somers takes a multivitamin. Oprah was convinced. "Many people write Suzanne off as a quackadoo," she said. "But she just might be a pioneer."

At the end of the day, Suzanne Somers feels like a different woman—a younger, healthier woman. "It has been four years now, and I'm feeling like a thirty-year old," wrote the sixty-year-old Somers in *Sexy Forever.* "I now realize this is the secret elixir we have all been looking for. People are always saying to me, 'You look great,' and I can see them studying my face. Best of all, my sex drive is back with a vengeance. I'm in the mood for love. It's so great at this age, after thirty-five years of marriage, to look at my husband and feel all 'wiggly' inside. And is he ever happy!"

Other celebrities have embraced Somers's regimen, including Simon Cowell. In 2001, Cowell claimed that an intravenous cocktail of vitamins B_{12}, C, and magnesium made him look and feel younger. "It's an incredibly warm feeling," said

Cowell. "You feel all the vitamins going through you. It's very calming."

Experts on aging haven't supported Somers's anti-aging revolution. In 2002, fifty-one of them, led by Jay Olshansky, Leonard Hayflick, and Bruce Carnes, weighed in. Olshansky is a professor in the School of Public Health at the University of Illinois and the author of *The Quest for Immortality: Science at the Frontiers of Aging.* Hayflick is a professor of anatomy at the University of California at San Francisco School of Medicine and the author of *How and Why We Age.* Carnes is a professor in the department of geriatric medicine at the University of Oklahoma Health Sciences Center. "No currently marketed intervention—none—has yet proven to slow, stop, or reverse human aging," they wrote. "Anyone purporting to offer an anti-aging product today is either mistaken or lying. Systematic investigations into aging and its modification are in progress and could one day provide methods to slow our inevitable decline and extend health and longevity. That day, however, has not arrived."

Somers doesn't take these criticisms lightly. Seeing a conspiracy among greedy pharmaceutical companies and uneducated, brainwashed doctors, she wrote, "In medical school the students receive very little instruction in endocrinology, and only four hours in how to prescribe hormones. If a doctor isn't curious, then his or her information comes primarily from the drug companies themselves. It doesn't take a rocket scientist to figure out that the information doctors get in a monthly throwaway magazine from the pharmaceutical companies would most likely be slanted; it is, after all, a business."

There is one thing, however, that Suzanne Somers is right about: we do live longer than we used to. And it's because she and many others offer advice like eat lots of fruits and vegetables, exercise, get plenty of sleep, don't smoke, avoid sugar, and reduce stress. People don't live longer because they've changed the way they *age*; they live longer because they've changed the way they *live*. But when Somers claims to slow or reverse the aging process, she enters a world of fantasy. She's not the first. Both Alexander the Great and Ponce de León searched for the legendary Fountain of Youth; and celebrities and healers posing as experts have been touting their magic elixirs ever since. It's an easy market. Everyone wants to live longer. "I don't want to achieve immortality through my work," said Woody Allen. "I want to achieve it through not dying."

Today's hucksters are no different from those found at sideshows a hundred years ago. Like Somers, they claim the only reason their therapies haven't entered the mainstream is that Big Pharma doesn't want them to. "The reason for the continued use of synthetic hormones," writes Christiane Northrup, "is that naturally occurring compounds cannot be patented. Therefore, using them has not been in the financial interest of drug companies." Somers and Northrup cast themselves in the same role: David versus Goliath. They're the little guys trying to help people stay young, while drug companies are the evil giants interested only in profit. Promoters of anti-aging medicines, through their websites, DVDs, books, and pamphlets, invariably advertise their products using a phrase they know will work: "what the pharmaceutical companies don't want you to know."

The irony is inescapable. For one, the anti-aging business has profits rivaling those of many pharmaceutical companies, making a fortune for its promoters. Suzanne Somers is an industry. On her website, she promotes only one brand of vitamins, supplements, and minerals: RestoreLife. There's RestoreLife Formula Essential Mineral Packets, Supplement Starter Kit, Resveratrol, Omega-3, and Vitamin D3, as well as RestoreLife Digest Renew, Bone Renew, Calm Renew, Natural Sleep Renew, and Sexy Leg Renew. Somers sells her own brand of foods, cooking utensils, and sweeteners (SomerSweet), as well as skin-care, weight-loss, and detoxification products. She sells nanotechnology patches to control appetite. All these products have made Suzanne Somers a multimillionaire. She's in the anti-aging business. And so are the doctors and compounding pharmacies she promotes in her books and on her website.

Although anti-aging gurus rail against mainstream medicine for not being on their side, their biggest problem is that science isn't on their side.

Olshansky, Hayflick, and Carnes argue that the biggest reason we age is oxidation, which releases free radicals that damage DNA. As DNA mutations accumulate, cell functions are impaired, causing an increased vulnerability to infection and disease. At the heart of the problem are mitochondria, small organelles in every cell that release free radicals while converting nutrients to energy. Because converting nutrients to energy is necessary for life—and because that process produces the free radicals that eventually kill us—we are, in effect, born to die. "It is an inescapable biological reality," they wrote, "that once the engine of life switches on, the body inevitably sows the seeds of its own destruction."

Olshansky, Hayflick, and Carnes published their critique of anti-aging medicines in *Scientific American* in 2002. At the time, they knew that supplemental antioxidants like selenium, beta-carotene, and vitamins A, C, and E had been proposed to counter the damaging effects of free radicals. Although studies of antioxidants were just getting started, and they didn't yet know the results, what they wrote was an ominous predictor of the future: "Antioxidants constitute one popular class of supplements touted to have anti-aging powers. Proponents claim that if taken in sufficient quantities, antioxidant supplements will sop up the radicals and slow down or stop the processes responsible for aging. But eliminating all free radicals would kill us, because they perform certain necessary intermediary steps in biochemical reactions." And that's exactly what happened. Studies have now shown that people who take large quantities of vitamins and dietary supplements with antioxidant activity are more likely to have cancer and heart disease and die sooner. "People might try a putative anti-aging intervention thinking they have little to lose," they wrote. "They should think again."

Free radicals aren't the only reason we age. In the early 1960s, Leonard Hayflick, then a scientist at the Wistar Institute, in Philadelphia, received fetal cells from an elective abortion performed in Sweden. Hayflick took the cells and bathed them in nutrient fluid in his laboratory. He wanted to see how often the cells would reproduce. What he found surprised him. No matter how attentive he was—no matter how many growth-promoting substances he put into the nutrient fluid—cells reproduced about fifty times before dying. Leonard Hayflick had proved what German biologist August Weissmann had postulated eighty years earlier: "Death takes place because cell division is not everlasting but finite."

Although the relative contributions of oxidation and limited cell division to mortality are unclear, one thing is certain: Suzanne Somers's herbs, coffee enemas, and glutathione liver rubs don't address the fundamental reasons for how and why we age.

Somers has written many books, with her picture on every cover. She's beautiful. In fact, she doesn't look any older than she did when she played Chrissy Snow on *Three's Company*. Remarkable, given that she was in her thirties then and is in her sixties now. But pictures can be deceiving. And because Somers's anti-aging medicines have no hope of reversing or slowing the aging process—and because she's in the business of saying they do—she has no choice but to resort to Plan B. On October 14, 2006, Somers appeared on *Larry King Live* to promote bioidentical hormones.

> KING: In addition to feeling good inside, do you look better outside?
>
> SOMERS: I ask you. Do I look better outside?
>
> KING: But you could have had work done. And I wouldn't know that.
>
> SOMERS: No. This is a real face. This is a hormone face.
>
> KING: You have not had plastic surgery?
>
> SOMERS: I have had some fillers.
>
> KING: What do you mean? Botox?
>
> SOMERS: Yes. Yes. Everybody does that.

"Today, we have available to us new techniques for youthfulness such as fillers like collagen and Botox," writes Somers. "The face-lifts of old look strange and outdated, and today's advantages used in moderation can help you maintain a youthful appearance without looking 'strange.' The object is to look natural." And if Botox and collagen don't work, Somers suggests shocking your face with electrical currents. "I have a thing called a FaceMaster," she told Larry King, "which I have been using for fourteen years. I hate to be self-serving, but I sell it on suzannesomers.com. It's a microcurrent face-lift machine . . . and it pumps up the muscles under your skin."

So, after all that, after taking dozens of vitamin, supplement, mineral, and herb pills every day, after rubbing estrogen and progesterone on her arms and glutathione over her liver, after injecting hormones and coffee into unnatural places, Suzanne Somers resorts to the one thing that can actually make her look younger: Botox. A direct contradiction to everything she's been preaching. It's hard to make the case that people should live naturally when you're injecting one of the most powerful toxins known to man (botulinum toxin) directly into your face. (Botulinum toxin is so powerful that as little as 0.00000001 grams can paralyze facial muscles.)

In February 2011, Somers's story took another bizarre turn. During an appearance on a Canadian talk show, fans noticed that her appearance had changed dramatically. "Suzanne's face looks very puffy and her lips look like sausages," said Tony Youn, a plastic surgeon in Detroit who had viewed pictures of Somers. "Those are the telltale signs of a stem-cell face-lift, in which doctors inject fat and stem cells under the skin." Stem-cell face-lifts are not approved in the United

States. In 2012, Somers used stem cells to reconstruct her breasts.

I n a way, it's all kind of sad—our unwillingness to accept getting older. "Anyone who has not been buried in a vault for the past two decades is surely aware of the media blitz touting the 'new old age' as a phenomenon that enables people in their sixties, seventies, eighties, nineties, and beyond to enjoy the kind of rich, full, healthy, adventurous, sexy, financially secure lives that their ancestors could never have imagined," wrote Susan Jacoby in *Never Say Die: The Myth and Marketing of the New Old Age*. "At eighty-five or ninety—whatever satisfactions may still lie ahead—only a fool or someone who has led an extraordinarily unhappy life can imagine that the best years are still to come."

Somers doesn't see it that way. "It is the year 2041," she wrote. "This is me, Suzanne Somers, at ninety-four years old. I am healthy, my bones are strong; my brain is working better than ever. I wake up happy, excited, and active. Most mornings start with wonderful sex with my one-hundred-and-five-year-old husband, Alan, who has also embraced the same health regimen. I am not one of those 'old people' put into a corner or, worse, in a nursing home. Nope, not me, I got it early on. I wanted to live, really live. So I jumped on the fast-moving train of the new medicine and never looked back. My friends laughed at me, called me a 'nut case' and a 'health freak,' but who's got the last laugh now?"

No one can deny Somers her optimism. No one can deny her an interest in living a better, fuller, more productive life.

But Suzanne Somers isn't just a citizen railing against the dying of the light. She's a paid promoter of a $6-billion-a-year anti-aging industry who hawks products that have no chance of helping and, because she includes megavitamins, every chance of hurting—a huckster who wants you to ignore the science. "It is not always easy, certainly from a nonscientist's perspective, to distinguish between real anti-aging science and the vast array of products, from unproven and untested supplements to self-help books by those who believe that age is just a number and a state of mind," writes Jacoby. "The last thing marketers want is for the public to make a clear-sighted, evidence-based assessment of whether such potions do anything more than enable denial of the physiological reality and inevitability of aging."

Suzanne Somers isn't the only celebrity to have created a cottage industry of alternative therapies. There is another television and movie star who believed she had found a cure for something the medical establishment had ignored. This time, however, the target audience wasn't adults with menopause or advancing age; it was parents desperate to find a cure for their children.

6

Autism's Pied Piper:

Jenny McCarthy's Crusade

When you think about it, what other choice is there but to hope?
—Lance Armstrong

Jenny McCarthy's film credits include *The Stupids*, *BASEketball*, *John Tucker Must Die*, and *Dirty Love*, which she also wrote. More recently, McCarthy has made guest appearances on *My Name Is Earl*, *Chuck*, *Just Shoot Me!* and *Two and a Half Men*. Her latest book, published in 2012, was *Bad Habits: Confessions of a Recovering Catholic*.

On September 24, 2008, Oprah interviewed McCarthy about her book *Mother Warriors: A Nation of Parents Healing Autism Against All Odds*. McCarthy's son, Evan, had been diagnosed with autism. Like Somers, McCarthy didn't trust mainstream doctors. They didn't know what caused autism or how to cure it. McCarthy, on the other hand, knew both. And she was there to tell mothers that it was time to take control. Time to be their own doctor. Oprah agreed. "During a production meeting not long ago," said Oprah, "one of my producers brought in an unforgettable article from the *Boston Globe Magazine* about the most extraordinary woman I've ever heard of. Right then and there we knew that she was somebody that we had to share with you on our show. Call your friends right now, because this woman, she's not just a mom—she's a warrior." Where doctors failed, Jenny McCarthy and Oprah Winfrey would succeed. And another counterfeit industry was born.

I n 1973, Bernard Rimland, a researcher at the Institute for Child Behavior Research, in San Diego, and the father of an autistic son, wrote a chapter titled "High Dosage Levels of Certain Vitamins in the Treatment of Children with Severe Mental Disorders." (The book was edited by Linus Pauling.) Rimland believed that large doses of vitamins and minerals could treat autism. He later founded the Autism Research Institute, which spawned Defeat Autism Now (DAN)—a group of clinicians dedicated to the notion that autism could be cured with vitamins and supplements. Where Somers aligned herself with gynecologist Christiane Northrup to promote bioidentical hormones, McCarthy aligned herself with Jerry Kartzinel, a DAN doctor, to promote treatments for autism.

In 2010, McCarthy and Kartzinel published a best-selling book titled *Healing and Preventing Autism: A Complete Guide*. Although McCarthy didn't start the movement to treat autism with biomedical therapies, she did, with the help of Oprah

Winfrey, bring it into the homes of tens of millions of Americans. McCarthy, Kartzinel, and doctors affiliated with DAN believed that autism had many causes and many cures. DAN doctors variously argued that:

- Autism is caused by mitochondrial dysfunction and should be treated with megadoses of vitamins A, C, D, E, K, and the B group, as well as zinc, selenium, calcium, magnesium, chromium, cod liver oil, omega-3 fatty acids, taurine, glutamine, arginine, creatine, carnitine, and coenzyme Q10.

- Autism is caused by food allergies and should be treated by restricting gluten (grains) and casein (dairy). "I started it," said McCarthy. "In two to three weeks Evan doubled his language."

- Autism is caused by overgrowth of fungi in the intestine and should be treated with antifungals and cow colostrum. "Once you detox that, these kids are getting better," said McCarthy. "You're cleaning up the gut. You're cleaning up the brain. There's a connection."

- Autism is caused by heavy-metal poisoning and should be treated with detoxifying therapies such as coffee enemas and intravenous ethylenediaminetetraacetic acid (EDTA). (In 2005, a five-year-old with autism named Tariq Nadama died of a heart arrhythmia after an intravenous injection of EDTA.)

- Autism is caused by misalignment of the spine and should be treated with vigorous chiropractic manipulations of the head and neck.

- Autism is caused by inflammation of the brain and

should be treated with *Curcuma longa*, a plant from the ginger family.

- Autism is caused by improper digestion of food and should be treated with digestive enzymes. "If our immune system is operating from our gut," wrote McCarthy, "how can it possibly do its job if it's filled with poop."

- Autism is caused by incorrect wiring of the brain and should be treated with electrical or magnetic stimulation.

- Autism is caused by an imbalance of immune cells and should be treated by infecting children with hookworms and whipworms.

- Autism is caused by a lack of oxygen to the brain and should be treated by placing children in hyperbaric oxygen chambers. (On May 1, 2009, a four-year-old boy with cerebral palsy, Francesco Martinizi, died after an explosion in a hyperbaric oxygen chamber caused burns over 90 percent of his body.)

- Autism is caused by a leaky gut and should be treated with probiotics.

- Autism is caused by immune dysregulation and should be treated with intravenous immunoglobulins or stem-cell transplantation.

- Autism is similar to a drug addiction and should be treated with low-dose naltrexone suspended in emu oil. (Naltrexone is used to treat drug dependency.)

- Autism is caused by excessive stimulation and should be treated with marijuana or melatonin.

- Autism is caused by a defect in metabolism and should be treated with shots of vitamin B_{12}. "It happened with

Evan," wrote McCarthy. "He was at UCLA autism school at the time and they said, 'What did you just do? He just had a burst of language.' And I said, 'B$_{12}$ shots.'"

- Autism is caused by chronic viral infections such as herpes and should be treated with antiviral medicines.

- Autism is caused by a blockage of the lymph glands and should be treated with lymphatic drainage massage.

- Autism is caused by intestinal parasites and should be treated with chlorine dioxide, a potent bleach used for stripping textiles and purifying industrial waste. (Bleach cocktails or enemas, which can be given as frequently as every two hours for three days, have caused severe vomiting and diarrhea.) Although one can only have sympathy for parents desperate to help their children, desperation can become child abuse.

- Autism is caused by vaccines. "Right before my son got the MMR [measles-mumps-rubella] shot, I said to the doctor, 'I have a very bad feeling about this shot. This is the autism shot, isn't it?'" McCarthy told Oprah. "And then the nurse gave [my son] that shot. And I remember going, 'Oh, God, no!' And soon thereafter I noticed a change. The soul was gone from his eyes." McCarthy didn't want other parents to make the same mistake, later writing, "Many people ask me if I had to do it all over again with a new baby, would I vaccinate? The answer is no. Hell no." Kartzinel, McCarthy's co-author, agrees, writing that children shouldn't receive vaccines if they have ever experienced cradle cap, constipation, diarrhea, sleep issues, tantrums, reclusiveness, transition issues, or red cheeks (in other words, everyone).

Oprah was impressed. Impressed that Jenny had written a book that contained so much good advice. Impressed that Jenny had become an expert in the treatment of autism. "She wrote the book," said Oprah. "She knows what she's talking about."

The vitamins, minerals, supplements, coffee enemas, and herbs recommended by McCarthy to treat autism are the same therapies that were recommended by Michael Schachter to treat Joey Hofbauer's Hodgkin's disease, William Kelley to treat Steve McQueen's mesothelioma, and Suzanne Somers to counter menopause and aging. Vastly different problems, eerily similar treatments.

Although McCarthy doesn't mention it in her books or television appearances, researchers have shed a great deal of light on the cause or causes of autism. For example, Ami Klin, at the Yale Child Study Center, studied babies who were only a few weeks old. He wanted to see how they attended to their mother's face, finding that those who were developmentally normal looked into their mother's eyes, while those later diagnosed with autism watched their mother's mouth. Eric Courchesne, at the University of California at San Diego, found structural abnormalities in the brains of children later diagnosed with autism when they were still in the womb. And Hakon Hakonarson, of the Children's Hospital of Philadelphia, along with many other investigators, found certain genetic abnormalities in autistic children. Researchers have also found that environmental factors can influence the risk of autism in the developing fetus—specifically, drugs like valproic acid (an anti-seizure medicine). Of interest, susceptibility to environmental influences appears to occur *before* children are born, not after.

Given our current understanding of the disorder, McCarthy's advice to treat autism as if it's caused by parasitic infections, heavy-metal poisoning, or blocked lymph glands is nonsense. So it shouldn't be surprising that whenever her therapies have been tested, they haven't worked. Worse: McCarthy's advice to avoid vaccines is not only useless; it's dangerous. Parents who choose not to vaccinate aren't lessening their children's risk of autism; they're only increasing their risk of suffering preventable diseases.

S ometimes it's hard to have much sympathy for the buyer. Adults who spend hundreds or thousands or even tens of thousands of dollars on the endless array of vanity items found in the cabinets of anti-aging gurus—all with the hope of turning back the clock—are going in with their eyes open. But when alternative healers take advantage of desperate parents, it's a different story. Parents of children with autism will do anything to help their children. Perhaps no story shows just how desperate parents can be than one involving an obscure intestinal hormone called secretin.

In the late 1990s, secretin became all the rage when a woman named Victoria Beck said that it had caused a dramatic improvement in her autistic son's language acquisition. Others also claimed remarkable results. So autism researchers decided to test it. They divided children into two groups; one received intravenous secretin, the other intravenous salt water. None of the parents knew which preparation their children had received. The results were interesting. Most parents in the secretin group rated their children as improving. But so did parents whose children had received salt water. In other words, parents had such a strong desire to see results following an expensive intravenous medicine that they believed their children were improving, regardless of what they'd received. It's hard to know why this was true. Maybe parents perceived children as better even though they weren't. Or maybe parents had become more attentive, causing them to appreciate subtle differences they hadn't seen before. Whatever the reason, salt water doesn't treat autism, so something other than the pharmacological effect of secretin had been at work. Fifteen studies have now shown that secretin is no better than a placebo for autism.

The most amazing part of the secretin story was what happened next. When parents were told that responses to secretin and salt water were indistinguishable, 69 percent still wanted to use the drug—still wanted to pay thousands of dollars and travel hundreds of miles to get something they now knew didn't work. That's how desperate they were. Because mainstream medicine didn't have anything better to offer—didn't have medicines that could make autism melt away—parents mortgaged houses and cashed in retirement accounts to find anyone who could promise hope, even if it was false hope. And even when they knew it was false.

Alison Singer, founder of the Autism Science Foundation and a graduate of Yale University and Harvard Business School, explains how easily well-educated parents can be duped. "When my daughter, Jodie, was diagnosed with autism,

I wanted to fix her," she said. "I wanted to do everything possible to make her better. What kind of mother would I be if I didn't try? At that point I didn't realize how lifelong autism would be. We tried gluten- and casein-free diets. We tried dimethylglycine. People said you had to sprinkle it on French toast. So I learned to make French toast." Singer's moment of clarity came when she saw a doctor who had been recommended by a friend. "One time I took Jodie to a chiropractor," she said. "He told me he could cure Jodie by rearranging the ions in her brain with a giant electromagnet placed under her mattress at night. And, 'oh, by the way,' he sells the magnets for two hundred dollars. So I went home and I talked to my husband about it. At this point, I had stopped being a smart person. And he just looked at me and said, 'Listen to yourself. Do you hear what you're saying?' It was that moment when I realized how far I'd gone. This was my grief, not my brain. And you can't think with your grief."

For Singer, acceptance of her daughter's disorder came slowly. "I was convinced that I was never going to be happy again until Jodie was cured. And I believed with all of my soul that it was just a matter of finding the right cure and then she was going to be fine. And then we would all be fine. And slowly I came to realize that it's a developmental disorder and that she's going to have challenges her entire life. When she was born, I looked into her beautiful little eyes and thought about her future and all the things we were going to do and all the joy we were going to have. But our lives have been very different. And it took a long time until I accepted that and was able to go back to thinking with my brain. That's when I finally sought out science-based interventions instead of quackery."

Singer doesn't blame parents. "I think the culpability lies with the quacks who are preying on the desperation of families. I think that's the worst kind of person who would take advantage of a parent or child during a time when they're grieving. I don't blame parents for being susceptible to this. I don't blame them for wanting to believe. You just can't imagine that there is someone who wants to take advantage of you."

I n the 1995 movie *The American President*, Lewis Rothschild pleads with President Andrew Shepherd to counter the attacks of his rival, Bob Rumson. Rothschild is angry that Rumson is the only one offering answers to America's problems.

SHEPHERD: Look, if people want to listen to [Rumson] . . .

ROTHSCHILD: They don't have a choice! Bob Rumson is the only one doing the talking! People want leadership, Mr. President, and in the absence of genuine leadership, they'll listen to anyone who steps up to the microphone. They want leadership. They're so thirsty for it they'll crawl through the desert toward a mirage, and when they discover there's no water, they'll drink the sand.

SHEPHERD: Lewis, we've had presidents who were beloved who couldn't find a coherent sentence with two hands and a flashlight. People don't drink the sand because they're thirsty. They drink the sand because they don't know the difference.

Alternative healers who promote secretin or spinal manipulations or hyperbaric oxygen chambers or ion-rearranging machines to treat autism are selling sand. They do it because of a misguided belief that their therapies work; they do it because it's lucrative; they do it because responsible advocacy organizations haven't stepped forward; and they do it because some parents don't know—or prefer not to know—the difference. Nothing is more contemptible than a clinician who takes advantage of loving parents by raiding their life savings.

F or children with diseases like diabetes, bacterial meningitis, and lymphoma, medicine offers cures like insulin, antibiotics, and chemotherapy. Not so with autism. McCarthy's treatments are seductive, in part, because medicine offers so little. (In her books, McCarthy promotes 260 chiropractors, naturopaths, dentists, doctors, and nurses who sell autism cures.) But the problem with McCarthy's campaign isn't only that her therapies don't work; it's that they might do harm. Children have died from medicines that bind heavy metals or suffered perforated eardrums in hyperbaric oxygen chambers or bone thinning from casein-free diets. Perhaps worst of all are the children who have suffered from McCarthy's very public denouncement of vaccines.

Before vaccines, Americans could expect that every year diphtheria would kill fifteen thousand people, mostly young children; rubella (German measles) would cause as many as twenty thousand babies to be born blind, deaf, or mentally disabled; polio would permanently paralyze fifteen thousand children and kill a thousand; mumps would be a common cause of deafness; and a bacterium called *Haemophilus influenzae* type b (Hib) would cause hundreds of children to die of suffocation from epiglottitis—no different than being smothered by a pillow. In the developed world, vaccines have completely or virtually eliminated these diseases.

Although acupuncturists, chiropractors, naturopaths, and homeopaths all come from different places in history—and offer therapies based on different philosophies—the one place they all seem to come together is vaccines, which they uniformly disdain. It's hard to know why. Maybe it's because it distinguishes them from their competition (mainstream doctors). Or maybe it's because they think vaccines are unnatural (although it's hard to make a case that coffee enemas are natural). Or maybe it's all part of the countercultural playbook (you're either on the bus or off the bus). Whatever the reason, it's done a lot of harm. And although most alternative healers don't have much national appeal, Oprah does. And when Oprah gave credence to McCarthy's anti-vaccine message, it had an effect. During the past few years, Americans have witnessed an increase in hospitalizations and deaths from diseases like whooping cough, measles, mumps, and bacterial meningitis, because some parents have become more frightened by vaccines than by the diseases they prevent.

Chronic Lyme Disease:

The Blumenthal Affair

It is absurd that the administration of a modern state should be
left to men ignorant of science.
—Frederick Soddy, British chemist

Movie and television stars aren't the only celebrities offering
medical advice. Politicians have also weighed in.

On February 11, 2009, Lawrence Gostin and John Kraemer, law professors at Georgetown University, published a
paper in the *Journal of the American Medical Association*. Typically, *JAMA* publishes papers written by doctors and scientists, not lawyers. But this was an unusual case. "Medical
science," they wrote, "and the health of patients who depend
on it, are too important to be subjected to political ideolo-
gies." Gostin and Kraemer were referring to the inexplicable
actions of Richard Blumenthal, attorney general of Con-

necticut. Blumenthal had tried to bully a medical society into creating a disease.

It wasn't the first time a politician had politicized science.

n 1977, Dan Burton, a Republican congressman from Indiana, stood proudly on the steps of the state capitol to announce that Indiana citizens should ignore FDA warnings and use laetrile as they pleased. Ten years later, Burton disagreed with the FDA again, this time for banning ephedra, a weight-loss product that had caused psychosis, hallucinations, paranoia, depression, irregular heartbeats, and strokes in hundreds of people. One, a thirty-four-year-old man who had taken ephedra for ten days, had jumped out of a second-story window to escape imagined attackers. Another, Steve Bechler, a pitcher for the Baltimore Orioles, had died less than twenty-four hours after taking the drug. But Burton was adamant. He accused the FDA of "harboring a culture of intimidation and sometimes harassment against alternative cures."

Burton's ignorance wasn't limited to cancer and weight control. When AIDS began spreading across the United States in the 1980s, he became obsessed with the disease, bringing his own scissors to the barbershop and refusing to eat soup in restaurants because he was unsure who was preparing his food. Later, he introduced (unsuccessful) legislation mandating HIV testing for every American.

But Dan Burton's greatest contribution to the science of the absurd came in the early 2000s, when he sponsored a series of congressional hearings that offered a platform to Andrew Wakefield, a British surgeon who had claimed that the measles-mumps-rubella vaccine caused autism. Wakefield's star didn't shine long. First, study after study failed to confirm his theory. Then a British journalist named Brian Deer found that Wakefield had received £440,000 from a legal services commission (a conflict of interest Wakefield had neglected to mention to his co-authors) and that he had misrepresented some of his clinical and biological data (causing the journal to retract the paper). Eventually, Andrew Wakefield was struck off the medical register in the United Kingdom—no longer able to practice medicine. But during Wakefield's fall, Burton never relented, continuing his public assault on MMR. As a consequence, parents of more than a hundred thousand American children chose not to give the vaccine. The results were predictable. In 2008, measles outbreaks were greater than in any year in more than a decade. In Europe, where Wakefield's claims stoked similar fears, thousands of children were infected, and at least thirteen died from measles, a preventable illness.

Incidentally, Burton was carrying on a time-honored Indiana tradition of trying to legislate bad science. On January 18, 1897, Indiana state representative Taylor I. Record argued in favor of changing the value of *pi*. *Pi*, which can be rounded to 3.14159, is the ratio of a circle's circumference to its diameter. Tyler believed that the number was inconveniently long; in House Bill 246, he asked that it be rounded up to 3.2. The bill passed the House but was defeated in the Senate when the chairman of Purdue University's math department successfully pleaded that it would make Indiana a national laughingstock. The value of *pi* in Indiana remains the same as in every other state.

B ut it was Attorney General Richard Blumenthal who took political shenanigans to a new level. Blumenthal tried to legislate a disease, Chronic Lyme, into existence.

In November 1975, Polly Murray, a mother of four living in Old Lyme, Connecticut, called the state health department to report twelve children who had suddenly suffered swelling, redness, and tenderness in their joints (arthritis). All lived in her small community of five thousand people, four on the same road. The doctors said it was juvenile rheumatoid arthritis (JRA), an autoimmune disease caused by the body's reacting against itself. To Polly Murray, this didn't make sense: how could an autoimmune disease cause an outbreak?

Polly wasn't alone. Later, another mother from the same community called the Yale rheumatology clinic to report that she, her husband, two of her children, and several neighbors had suddenly developed arthritis. Again, all were said to have JRA.

The task of deciphering the events in Old Lyme fell to a young postdoctoral fellow in Yale School of Medicine's rheumatology division, Allen Steere. Steere studied fifty-one victims of the disease—thirty-nine of whom were children. He agreed with Polly Murray: not in keeping with JRA, these cases were seasonal, involved only one joint, were associated with an unusual rash, and occurred in an unlikely number of people in one town in one summer, 10 percent of whom lived on one of four roads. Given the prevalence of JRA in the United States, the chance of that happening was 100 to 1.

In January 1977, Steere and his coworkers published a paper that gave the disease its name: "Lyme Arthritis." Steere didn't know what was causing the disease, but he had a sense of how it was transmitted: "The geographical clustering of the patients in more sparsely settled, heavily wooded areas rather than in town centers or along the shore [and] the peak occurrence in summer months are best explained by transmission of an agent by an arthropod vector." The most common arthropods in the woods of Old Lyme are ticks, fleas, spiders, and mosquitoes.

Five years later, a bacteriologist named Willy Burgdorfer figured out what was happening in Old Lyme, Connecticut. Burgdorfer had studied in Basel, Switzerland, before coming to Hamilton, Montana, to work at the Rocky Mountain Laboratories, part of the United States Public Health Service. The Rocky Mountains are loaded with the same arthropod that would prove to be the cause of the Old Lyme outbreak: ticks. In 1982, Burgdorfer dissected the guts of *Ixodes* ticks and found corkscrew-like bacteria similar to those that caused syphilis. When Burgdorfer injected these bacteria into rabbits, they developed a rash identical to that found in Old Lyme. The bacteria were later called *Borrelia burgdorferi*.

With the bacteria that caused Lyme arthritis in hand, understanding the disease got a lot easier. Researchers could now develop methods to detect bacterial proteins and genes, enabling them to tell where the bacterium went, when it went there, how the body responded, and whether antibiotics worked. Within a few years it became clear what Lyme disease was. And what it wasn't.

Lyme disease occurs in three stages. First, ticks inject bacteria under the skin, where they reproduce and move outward, causing the characteristic bull's-eye rash: red, with heaped-up borders and central clearing. This first stage lasts days to weeks.

During the second stage, Lyme bacteria spread into the bloodstream and travel to other parts of the body. Patients might experience fatigue, fever, swelling of lymph nodes, more bull's-eye rashes, and neck, muscle, and joint pain (arthralgia). In about 15 percent of patients, the bacteria cause meningitis (inflammation of the lining of the brain and spinal cord) with neck stiffness and fever; encephalitis (inflammation of the brain itself) with headache and intolerance to light; facial palsy (also called Bell's palsy), in which one side of the face droops uncontrollably; and neuritis (inflammation of the nerves), causing pain, weakness, and numbness. In about 5 percent of patients, bacteria cause carditis (inflammation of the heart), disrupting electrical pathways necessary for normal heart rhythms; symptoms include fainting and chest pain. Interestingly, even without antibiotics, all of these symptoms usually resolve.

During the third stage, which occurs months after the tick bite, about 10 percent of untreated patients develop arthritis, mostly in large joints like the knee. In some untreated patients, arthritis persists or recurs.

When Allen Steere reported the outbreak in Old Lyme, he predicted that antibiotics wouldn't make a difference, impressed as he was that symptoms often disappeared without them. But Steere was wrong. Many studies have now proved that people treated with antibiotics resolve their symptoms more quickly and, if treated early, are less likely to progress to later stages. As a consequence, antibiotics are given by mouth or intravenously for two to four weeks.

I n Lyme disease, alternative healers saw an opportunity. Typically, they catered to those for whom medicine offered little, such as women with menopause or children with autism. Lyme allowed them to expand their repertoire. Claiming that Lyme had a fourth stage, which they called Chronic Lyme disease, they argued that Lyme bacteria were the real cause of diseases like autism, chronic fatigue syndrome, fibromyalgia, reflex sympathetic dystrophy, homicidal behavior ("Lyme rage"), birth defects, Parkinson's disease, multiple sclerosis, and Lou Gehrig's disease. Because only they knew of this fourth stage, they called themselves "Lyme Literate" doctors, warning patients that they might have Chronic Lyme if they suffered weight change, hair loss, sore throat, menstrual irregularity, upset stomach, constipation, diarrhea, cough, headache, neck pain, lightheadedness, motion sickness, poor balance, wooziness, tremor, confusion, difficulty concentrating, forgetfulness, mood swings, disturbed sleep, or hangovers after drinking alcohol.

The so-called Lyme Literate doctors created an industry. They wrote books like *Beating Lyme*, *The Lyme Disease Solution*, *Healing Lyme Disease Naturally*, *The Top 10 Lyme Disease Treatments*, and *Insights into Lyme Disease Treatment: 13 Lyme Literate Health Care Practitioners Share Their Healing Strategies*. They made a movie titled *Under Our Skin* that bravely told the story of Chronic Lyme disease and the fight for recognition. "What has gotten under our skin," said the director, "is not just a microorganism, but medicine itself." The movie featured a pathologist, Alan MacDonald, who, after setting up a laboratory in his own basement, found the long-lived bacteria he believed caused Chronic Lyme, trapped in an ominous biofilm. It featured doctors like Joseph Burrascano, from East Hampton, New York, and Joseph Jemsek, from Charlotte, North Carolina, who had devoted their lives to treating Chronic Lyme. And it told stories of patients who, after receiving a variety of alternative therapies, had recovered from debilitating pain and fatigue. The movie also included uncaring mainstream doctors—supposed experts in Lyme disease—who denied that Chronic Lyme even existed, passing off sufferers as cranks, malingerers, or psychiatric patients. It wasn't hard to tell the good guys (Lyme Literate doctors and their patients) from the bad guys (everyone else). *Under Our Skin* won awards at several film festivals and was a nominee for Best Documentary Feature at the Academy Awards. The movie also hinted at how deeply alternative healers had reached into their grab bag, with therapies ranging from the predictable to the unimaginable. Lyme Literate doctors offered:

- Vitamins such as A, B_{12}, C, and D given in high doses by mouth or intravenously.
- Supplements such as beta-carotene, glutathione, glutamine, 5-hydroxytryptophan, alpha-lipoic acid, chromium, magnesium, coenzyme Q10, omega-3 oils, inositol, gamma-aminobutyric acid, L-threonine, linoleic acid, grapeseed extract, folic acid, zinc, chlorella, digestive enzymes, and probiotics.
- Herbs and other natural products such as turmeric,

ginger, garlic, curcumin, astragalus, aloe vera, cat's claw, Japanese knotweed, andrographis, *Stephania* root, devil's claw, white willow bark, *Boswellia*, kava, green tea, St. John's wort, valerian root, slippery elm, kefir, marshmallow root, licorice root, bovine colostrum, olive leaf extract, sarsaparilla, lauricidin, Siberian ginseng, silicon, resveratrol, melatonin, nettle, capsaicin cream, bromelain, humperzine, vinpocetine, carnitine, periwinkle, hawthorn, khella, red root tincture, capryl, and passion flower.

- Techniques such as hyperbaric oxygen, intravenous hydrogen peroxide, acupuncture, magnets, enemas, sweat lodges, lymphatic drainage massage, intravenous chelation, laser energy detoxification, reverse spin therapy, biophoton therapy, ozone saunas, and emu oil.

- Homeopathic remedies such as arnica.

- And diets that prohibit gluten and casein.

The use of vitamins, supplements, herbs, diets, homeopathy, saunas, oils, chelation, acupuncture, and magnets are typical fare from the alternative medicine menu. But Lyme Literate doctors have shown a level of imagination well beyond that of their colleagues. Wolf Storl claims that Chronic Lyme can be treated with teasel, a common roadside weed (*Dipsacus sylvestris*). In his book, Storl gives specific instructions on how to effect the teasel cure, recognizing that it's not just the plant but the plant's spirit that's important. (According to Storl, teasel is affiliated with the planet Mars.) "One takes time with the plant," he writes, "sits down facing east, the direction of the rising sun, and opens all one's senses regarding it. Before getting into this meditation, one can burn dried sacred herbs—prairie sage or mugwort, for example—and smudge oneself. After contacting the plant spirit and asking for its help, one can dig out the root or harvest the leaves." Storl's book is packed with testimonials. "I take three full tablespoons a day," writes Dirk, "which is wonderfully bitter like medicine should be. Now, no more pain." The teasel cure isn't limited to people; dogs and Arabian horses have also apparently benefited.

Another imaginative treatment for Chronic Lyme is the Rife machine, invented by Royal Raymond Rife in the 1930s to treat tuberculosis. Bryan Rosner has written two books about Rife machines for Chronic Lyme. Rosner explains, "a Rife machine delivers . . . an invisible electromagnetic field to the body. The field passes through the entire body and disables targeted microorganisms." Rosner's books, like Storl's, are filled with glowing testimonials. "My daughter was bedridden for five months," writes Robin. "Using Rife machines, she won the state table tennis junior championship." In 2006, more than three hundred people attended a Rife machine convention in Seattle. These machines, which perform the supernatural task of electrocuting Lyme bacteria while leaving healthy cells intact, can be purchased on the Internet for between $400 and $2,500.

Although Lyme Literate doctors offer a wide range of therapies, no treatment receives more attention than long-term, high-dose intravenous antibiotics.

According to Lyme Literate doctors, Lyme bacteria hide. They hide in cells and in biofilms. They hide in the heart, brain, joints, and muscles. Indeed, they're apparently so well hidden that even the immune system can't find them and make antibodies. (The concept of Lyme disease without Lyme antibodies allows alternative healers to treat patients who never had the disease or who live in states where the disease doesn't exist.) Because Lyme bacteria are supposedly hidden, two to four weeks of antibiotics aren't enough; if patients really want to rid themselves of fatigue, pain, and memory loss, they need to be treated for months or even years. Then and only then will this terrible infection be gone.

Lyme Literate doctors' protests aside, a mountain of evidence fails to support their bacteria-are-hiding-in-the-body-but-you-just-can't-find-them claim:

First, Lyme bacteria are easily detected in the laboratory. When Lyme disease begins, bacteria are found in the expanding bull's-eye rash. After a couple of weeks of antibiotics, Lyme bacteria no longer grow from the rash (or anywhere else). Plus, unlike bacteria like chlamydia, mycoplasma, and rickettsia, Lyme bacteria don't grow inside cells. They grow outside cells, so they're easy to detect and kill. And unlike some other bacteria, Lyme bacteria haven't developed resistance to antibiotics. As a consequence, they are easily killed in laboratory flasks, in laboratory animals, and in people. The claim by Lyme Literate doctors that bacteria are hiding out of the sight of researchers (despite appropriate antibiotics) is akin to the claim by Bigfoot Literate people that only they know the monster exists.

Second, the notion that Lyme patients suffer long-term symptoms to a greater extent than the general population is also incorrect. Several studies have shown that Lyme sufferers don't develop chronic pain or fatigue more frequently than people who never had the disease. Sadly, one study showed that 50 percent of those diagnosed by alternative healers with Chronic Lyme had treatable disorders like depression, rheumatoid arthritis, bursitis, and myasthenia gravis (an autoimmune disease that affects the muscles).

Finally, the best way to determine whether Lyme Literate doctors are right is to take patients who they claim are suffering from Chronic Lyme and divide them into two groups. One group would be given long-term antibiotics, the other a placebo. This study has been done four times, each time with the same result. Patients treated with antibiotics for supposed Chronic Lyme fare no better than those given a placebo. Predictably, about a third of placebo recipients claimed that their symptoms had resolved.

In 2007, in response to overwhelming evidence that refuted the existence of Chronic Lyme—and clear evidence of harm caused by its treatment—a group of Lyme experts issued a definitive statement in the *New England Journal of Medicine.* "Chronic Lyme disease is a misnomer," they wrote, "and the use of prolonged, dangerous, and expensive antibiotic treatment for it is not warranted." In other words, long-term antibiotics for a disease that doesn't exist are unnecessary and dangerous.

To say the least:

In May 1999, a thirty-year-old woman entered the Mayo Clinic in Rochester, Minnesota, with jaundice and confusion. Despite aggressive attempts at resuscitation, she died soon after. Her diagnosis: a fatal blood clot that had lodged in her heart, a consequence of an indwelling intravenous catheter. Lyme Literate doctors had tested her urine, blood, and spinal fluid for the presence of Lyme bacteria and Lyme antibodies.

Every test had been negative. They treated her anyway. At the time of her death, she had received intravenous antibiotics for more than two years.

In 1993, the CDC, in conjunction with the New Jersey Department of Health, investigated an unusual outbreak in Monmouth and Ocean counties. Twenty-five people, mostly young girls, had had their gallbladders removed. All had been treated for months to years with intravenous ceftriaxone, an antibiotic known to cause gallstones. Although Lyme Literate doctors had diagnosed Chronic Lyme, most of these children didn't have any evidence of Lyme infection. Patients given long-term antibiotics for Chronic Lyme have also suffered antibiotic-resistant bacterial infections, severe allergic reactions, and bone marrow suppression.

Antibiotics aren't the only problem. In 1990, the New Jersey Department of Health investigated two cases of malaria in U.S. citizens; both were quite ill, both were infected with the same strain (*Plasmodium vivax*), and both had traveled to a clinic in Mexico to get infected, intentionally. (Lyme Literate doctors call this "fever therapy.") With proper anti-malaria therapy, both survived. In 2006, one person died and another was hospitalized after Lyme Literate doctors had recommended bismuth, a heavy metal. Lyme Literate doctors have been suspended, fined, reprimanded, or jailed for tax evasion, wire fraud, mail fraud, money laundering, insurance fraud, improperly prescribing antibiotics, using veterinary drugs in people, diagnosing children without examining them, and injecting patients with hydrogen peroxide and dinitrophenol, a toxic substance banned for medical use in the United States for fifty years.

B ecause so much harm had occurred in the name of Chronic Lyme, the Infectious Diseases Society of America (IDSA) decided to call out the cottage industry that had conjured up its existence and its treatments. In 2006, it issued guidelines to practitioners: "Because of a lack of biological plausibility, lack of efficacy, absence of supporting data, and the potential for harm to the patient, the following are not recommended for treatment of patients with [so-called Chronic] Lyme disease: hyperbaric oxygen, ozone, fever therapy, intravenous immuno-globulin, cholestyramine, intravenous hydrogen peroxide, spe-cific nutritional substances, vitamins, magnesium, and bismuth injections." It was one of the few times that a professional med-ical society had taken on practitioners of alternative medicine.

It almost proved their undoing.

W hen the IDSA published guidelines asking clinicians to avoid alternative medicines for a nonexistent disease, it threatened an industry. Lyme Literate doctors were making millions off the deception; they weren't about to take the IDSA's directives lying down. Indeed, a few years earlier, the Interna-tional Lyme and Associated Diseases Society (ILADS)—an organization composed of Lyme Literate doctors and their patients—had published its own guidelines advising exactly the opposite of what the IDSA had proposed. Three Chronic Lyme activist groups—the New Jersey–based Lyme Disease Association, the California Lyme Disease Association, and the

Connecticut-based Time for Lyme—lobbied state attorneys general hoping to find one to take up their cause. They found their champion in the state where Lyme had been born.

Richard Blumenthal was raised in Brooklyn, New York. After graduating from Harvard, he attended Yale Law School, where he was editor-in-chief of the *Yale Law Journal* and a classmate of Bill and Hillary Clinton. In 1990, he was elected Connecticut's attorney general, and he was reelected in 1994, 1998, 2002, and 2006. In Blumenthal, Lyme Literate doctors knew they had a friend. He had served on the advisory board of Time for Lyme, received awards from Chronic Lyme activist groups, drafted a state law assuring residents of long-term antibiotic therapy, criticized the state health department for underreporting Lyme disease, and supported a pediatrician, Dr. Charles Ray Jones, accused of prescribing antibiotics over the phone for two children in Nevada—a desert state with few ticks—without ever examining them. (Jones had pulled up to his hearing in a stretch limo, to the cheers of his many fans.)

In November 2006, one month after publication of its guidelines, Richard Blumenthal sued the IDSA for violating antitrust laws, claiming that it was an unauthorized monopoly. Angry that both the IDSA and the American Academy of Neurology (AAN) had, in his opinion, brainwashed clinicians, he argued that guidelines "improperly ignored or minimized consideration of alternative medical opinion and evidence regarding Chronic Lyme disease." Blumenthal claimed that the IDSA and AAN had "not only reached the same conclusions regarding the non-existence of Chronic Lyme disease, [but] their reasoning at times used strikingly similar language." To Richard Blumenthal, the IDSA and AAN were part of a larger conspiracy, and he wasn't about to let them get away with it. Attorneys for the AAN argued the obvious: the IDSA and AAN had reached the same conclusions because they had relied on the same evidence.

Unfortunately for Blumenthal, ten years earlier, both the Federal Trade Commission and the Department of Justice had ruled that treatment guidelines issued by medical societies didn't limit competition and, therefore, weren't subject to antitrust litigation. Blumenthal was a lawyer. He knew that if the law was on your side, you argued the law (the law wasn't on his side); if the facts were on your side, you argued the facts (IDSA guidelines was based on more than four hundred scientific studies); and if neither was on your side, you attacked the witness. So Blumenthal attacked the witness, claiming that five of the fourteen scientists who had drafted the IDSA guidelines hadn't disclosed financial relationships with drug and diagnostic companies. Given that the IDSA had recommended *against* unnecessary tests and treatments, Blumenthal's logic was hard to fathom. If followed, the IDSA guidelines would generate *less* revenue for drug and diagnostic companies, not more.

Indeed, if anything, the conflicts were on the other side. When ILADS issued guidelines friendly to Lyme Literate doctors, it failed to disclose that one committee member was the president of a company that sold an alternative Lyme diagnostic test. Furthermore, several state medical boards had disciplined Lyme Literate doctors for receiving payments directly from intravenous infusion companies. And, unlike the physicians and scientists who constituted the IDSA committee, Lyme Literate doctors had made a fortune treating Chronic Lyme. One practitioner, who practiced in a state that didn't have many patients with actual Lyme disease, collected $6 million from one insurance company in one year. (After his North Carolina practice declared bankruptcy, he opened a cash-only practice, spending $8 million on a building with a waterfall and a grand piano.) Another Chronic Lyme doctor, a former president of ILADS, is said to have treated two thousand Lyme patients in his San Francisco practice—impressive, given that between 1998 and 2007, fewer than a thousand Lyme cases were reported in California. (Before he entered the world of Chronic Lyme, he was an associate medical director at a penis-enlargement clinic.)

If the IDSA guidelines held, insurance companies might stop paying for expensive therapies; nothing threatened Lyme Literate doctors more than that.

B y May 2008 the IDSA had already spent $250,000 fending off Blumenthal's lawsuit. The IDSA was an organization of nine thousand members with meager dues—it didn't have much money. "One of the things that really worried us was that we were not a financially large organization compared to the Office of the Attorney General of the State of Connecticut," recalled Anne Gershon, president of IDSA at the time. "We were really worried that they would be able to break us financially." So both sides came to an agreement. The IDSA would form another committee. This time, however, Howard Brody, an ethicist from the University of Texas in Galveston, would determine who had unacceptable conflicts of interest.

On April 22, 2010, the new IDSA committee issued its final report. The panel had carefully reviewed the guidelines for more than a year, conducted a public hearing, taken evidence from more than 150 people, and issued a sixty-six-page document that contained 1,025 references. It is hard to imagine how they could have been more thorough. The results were the same. The new panel agreed with the previous panel that Chronic Lyme didn't exist, concluding, "the recommendations contained in the 2006 guidelines were medically and scientifically justified on the basis of all of the available evidence and no changes to the guidelines are necessary." Blumenthal had insisted that at least 75 percent of new panel members support the 2006 guidelines or they would be revised. In fact, 100 percent supported them. Lyme Literate doctors were outraged. They argued that Brody, by restricting panel members to those who had made less than $10,000 a year caring for Lyme patients, had excluded several panelists who believed in Chronic Lyme (and had made far more than that).

C hronic Lyme advocates had had their day in court, several testifying in front of the new panel. "It was a very collegial meeting," recalled Carol Baker, professor of pediatrics at Baylor College of Medicine and head of the new panel. "We were more credible and open-minded than they had expected. We talked to them at the breaks. We cared." Lyme organizations were so convinced that the guidelines would be revised in their favor that they celebrated in the same hotel where the panelists stayed. "I think they were expecting a different outcome," recalled Baker. When the panel upheld the previous recommendations, the Chronic Lyme community targeted Baker. "I got Internet threats that said, 'You have no compassion for patients,' 'You don't care about people.' The e-mails went on for a year. It wasn't fun."

After the second panel agreed with the first, Blumenthal said his office was "reviewing the IDSA's reassessment of its 2006 Lyme disease guidelines to determine whether [it had] fulfilled requirements of the settlement." And that was it. No apology. Nothing. "I never heard another word from him," recalled Baker. Certainly Blumenthal owed an apology—an apology to the little girls who had had their gallbladders removed, or the woman who had died from long-term antibiotic therapy, or the man who had died from bismuth therapy, or the two New Jersey residents who had suffered from malaria therapy, or the woman who had cashed in her retirement account to pay the $15,000 monthly fees required by Lyme Literate doctors to treat a disease that doesn't exist. A disease that Blumenthal had done everything he could to promote to the press and the public.

Richard Blumenthal had tried to make Chronic Lyme a story about people, not science. Although he might have eventually stacked the committee with enough fringe doctors to change the IDSA guidelines, he wouldn't have changed the science of Lyme disease any more than Dan Burton would have changed the facts about laetrile or Taylor Record the value of *pi*. In their critique of the Blumenthal affair, Gostin and Kraemer summed it up best. "The scientific process is not democratic," they wrote. Science isn't about who gets the most votes; it's about the quality, strength, and reproducibility of the evidence.

On January 5, 2011, Richard Blumenthal was sworn into the United States Senate, taking the seat vacated by Christopher Dodd. He was immediately appointed to the Health, Education, Labor and Pensions committee.

Part V

THE HOPE BUSINESS

Curing Cancer:

Steve Jobs, Shark Cartilage, Coffee Enemas, and More

> Of all the ghouls who feed on the bodies of the dead and dying, the cancer quacks are the most vicious and heartless.
>
> —Morris Fishbein, former editor of the *Journal of the American Medical Association*

In October 2003, Steve Jobs, cofounder of Apple Computer, was diagnosed with pancreatic cancer. A brilliant, restless innovator, he designed mp3 players, smartphones, and the first successful line of personal computers. Jobs liked being in charge, liked making decisions. Faced with cancer, he took control. Most pancreatic tumors are untreatable (so-called adenocarcinomas), but Jobs had a neuroendocrine tumor—with early surgery, he had an excellent chance of survival. Unfortunately, for nine months, Jobs, a Buddhist and vegetarian, treated himself with acupuncture, herbal remedies, bowel cleansings, and a special cancer diet consisting of carrot and fruit juices. His choice proved fatal. By the time he had surgery, the cancer had spread. On October 5, 2011, after spending years in and out of the hospital, Steve Jobs died of a treatable disease.

Jobs wasn't the first to be seduced by bogus cancer cures.

I n the 1800s, medicine men sold wondrous elixirs that—once analyzed—weren't so wondrous. Benjamin Bye's Soothing Balmy Oil cancer cure contained cottonseed oil, almond oil, sarsaparilla, talcum, and Vaseline. Chamlee's Cancer Cure— made from "an exotic Pacific Island shrub"—contained alcohol, iron, saccharine, and strychnine; and Curry's Cancer Cure contained hydrogen peroxide, iodine, laxatives, and cocaine. The Radio-Sulpho Cancer Cure had Epsom salts in a cast of Limburger cheese, netting $1 million for its inventor. The Toxo-Absorbent Cure contained sand and clay (to draw the cancer out). And Dr. Rupert Wells sold Radol, advertised as "radium-impregnated fluid," though it didn't contain any radium. Even John D. Rockefeller's father got in on the act, selling his bogus cancer cures at county fairs with the help of magicians, hypnotists, and ventriloquists.

Early American cancer cures were accompanied by heart-rending testimonials, were a gold mine for their promoters— and were utterly worthless.

I n the early twentieth century, Albert Abrams married America's fear of cancer to its love of gadgetry.

Abrams was born in San Francisco in 1864. After attending

the University of Heidelberg, in Germany, he returned to California to become a respected neurologist, professor of pathology, and vice president of the California State Medical Society. Then, in 1910, Albert Abrams became "the dean of twentieth-century charlatans."

Abrams claimed that cancers—as well as diseases like tuberculosis, gonorrhea, and syphilis—emitted different vibrations, like radio waves. To detect them, he invented the Dynamizer, a boxed jungle of coils, batteries, and rheostats. Two wires came out of the box: one plugged into a wall socket, and the other cupped onto the patients' forehead. To make the correct diagnosis, Abrams took a drop of the patients' blood and placed it inside the box. Patients then stripped to the waist, faced west, and stood in a dimly lit room while Abrams felt their abdomen. The Dynamizer could also detect the patients' place of birth, ethnic background, year of death, religion (Jews had duller abdomens than Christians), and golf handicap. Abrams leased his machine for $250 (the equivalent of about $8,000 today), with a $5 monthly user fee. Robert Millikan, winner of the Nobel Prize in Physics in 1923, described the Dynamizer as something "a ten-year-old boy would build to fool an eight-year-old."

Abrams soon realized that diagnoses alone weren't enough, so he invented the Oscilloclast, which sent out specific vibrations to counter disease. No one was more impressed than Upton Sinclair, the Pulitzer Prize–winning author of *The Jungle*. In an article titled "The House of Wonders," Sinclair praised Abrams's breakthrough technology, an endorsement that sent sales through the roof. By the early 1920s, thousands of doctors, mostly chiropractors, were using Abrams's machines. Like many scientists, however, the senior editor of *Scientific American* didn't buy Sinclair's endorsement, arguing that his name "meant no more in medical research than Jack Dempsey's would mean in a thesis dealing with the fourth dimension or Babe Ruth's on the mathematical theory of invariance."

Eventually, scientists put Abrams's Dynamizer to the test, sending him blood from animals and dead people. Abrams— assuming that he was evaluating living men and women— diagnosed a guinea pig with cancer, a sheep with syphilis, a rooster with sinusitis, and a dead man with colitis. The public never caught on to the fraud. When Abrams died in 1924, he had amassed a fortune of more than $2 million.

In the 1940s, William Koch invented a bogus cancer cure called glyoxylide, a combination of two carbon monoxide molecules. Unfortunately, carbon monoxide molecules don't stay together very long—separating in less than a hundred-millionth of a second into a gas. Koch sold his cure to thousands of doctors, who charged $300 per injection. When analyzed, chemists found that Koch's glyoxylide contained water—and water only.

In the 1950s, several years after researchers at Boston Children's Hospital had proved that chemotherapies like aminopterin could treat certain cancers, Harry Hoxsey invented yellow, pink, and black medicines. Yellow medicine was a paste made of arsenic to treat skin cancer. Pink and black medicines treated "internal" cancers. Pink medicine contained pepsin (a meat tenderizer) and potassium iodide (a chemical used to iodize table salt). Black medicine contained prickly ash bark, buckthorn bark, barberry root, licorice root, pokeweed, alfalfa, and red clover blossoms (together, a botanical laxative).

Among the twentieth century's greatest cancer frauds, no one was more successful than Hoxsey. After he had dropped out of school in the eighth grade, he supported himself as a coal miner, insurance salesman, and boxer. But it was an old family remedy that changed his life. Hoxsey's great-grandfather, a Kentucky farmer, claimed to have cured a horse of bone cancer with mashed flowers, ground stalks, and boiled roots. Harry's father inherited the formula and passed it on; black medicine was born.

Hoxsey moved around, selling his cancer cures in Iowa, Michigan, West Virginia, and New Jersey, often getting convicted of practicing medicine without a license. Eventually, he moved to Dallas and got a license as a naturopath, effectively keeping the police at bay. Within a few years, Hoxsey owned clinics in Denver, Los Angeles, and fifteen other cities. By the early 1950s—with ten thousand patients on the books—Hoxsey was earning $1.5 million a year. On his desk he had a sign that read, "The world is made up of two kinds of people—'dem that takes and 'dem that gets took." Hoxsey *took*, buying oil wells, real estate, automobiles, an airplane, and a six-hundred-acre ranch. "The man upstairs is smiling favorably on me," he said. "I'm a friend of Mr. H. L. Hunt, the wealthiest man in the nation, and he'd better look out, I'm catching up with him."

By the mid-1950s Harry Hoxsey's clinic in Texas was the largest private cancer clinic in the United States. Then Hoxsey confused commercial success with clinical success. Anxious to be taken seriously, he sent records of his most dramatic cures to scientists at the National Cancer Institute, who found that Hoxsey's patients had either never had cancer, still had cancer, were cured before they came to his office, or were dead.

Hoxsey's career ended precipitously when the parents of a sixteen-year-old boy with bone cancer chose pink and black medicines instead of a lifesaving amputation. After the family successfully sued him for fraud, Hoxsey fled to Tijuana and set up another clinic. In 1974, Harry Hoxsey died of prostate cancer complicated by a heart condition.

I n the 1960s, Andrew Ivy promoted the next great cancer cure: Krebiozen. Ivy's rise was matched only by his fall.

Andrew Ivy was born in Farmington, Missouri, in 1893. He attended the University of Chicago and Rush Medical College, where he received his bachelor of science in 1916, master of science in 1917, PhD in 1918, and MD in 1922. During the next thirty-five years, Ivy published fifteen hundred papers in the fields of gastroenterology, aviation medicine, reproductive medicine, blood clotting, artificial respiration, cardiac pain, flash burns, and typhoid infections. By 1970 Ivy's work was cited more than that of any other scientist in the world. His textbook, *Peptic Ulcer*, written in 1950, is still a classic.

In recognition of his achievements, Ivy was made head of physiology and pharmacology at Northwestern University Medical School and vice president of the University of Illinois. But to most Americans, Andrew Ivy was the man who helped prosecute Nazi doctors accused of torture and murder—and the author of "The Nuremberg Code," a manifesto on the ethics of human experimentation. (For his efforts, Ivy was awarded the

Certificate of Merit from the president of the United States.) Jonathan Moreno, in his book *Undue Risk: Secret State Experiments on Humans*, wrote "by the end of the war [Ivy] was probably the most famous doctor in the country."

Then, in July 1949, after a visit from Stevan Durovic, Andrew Ivy began his uninterrupted fall from grace. Durovic, who had fled his native Yugoslavia for Argentina, believed he had a cure for cancer. While in Argentina, Durovic injected horses with a bacterium called *Actinomyces bovis*, collected their serum, freeze-dried it, and resuspended it in mineral oil. He called his new medicine Krebiozen. When Durovic came to the United States, he knew that Andrew Ivy was the man to see. For years, Ivy had studied dogs with cancer of the thyroid, believing they produced a natural anti-cancer substance; but he couldn't find it. Durovic, on the other hand, believed he had.

Before coming to Ivy, Durovic had tested Krebiozen on twelve dogs and cats with cancer, claiming that all had improved or been cured. Ivy was impressed, so he injected himself, a colleague, and one dog with the drug. Convinced of its safety, Ivy injected the first cancer patient on August 20, 1949. On March 27, 1951, after treating twenty-two patients, he announced his findings at a press conference at the Drake Hotel, in Chicago, inviting local doctors, potential financial supporters, the mayor of Chicago, two United States senators, and science writers from four Chicago newspapers. Ivy claimed that all his patients had experienced "dramatic improvement." In truth, ten of the twenty-two were dead—all from cancer.

Ivy's international reputation spurred interest in Krebiozen. Within a few years more than four thousand cancer sufferers had been treated. In 1952, cancer specialists reviewed patient records and concluded that Krebiozen was worthless. In 1961, the National Cancer Institute, after conducting its own study, reached the same conclusion. When FDA chemists finally analyzed Krebiozen, they found it contained mineral oil and creatine, a harmless substance found in muscles. Krebiozen was a hoax.

In 1964, Andrew Ivy and Stevan Durovic were indicted on forty-nine counts of mail fraud, mislabeling, conspiracy, and making false statements. After he was acquitted, Durovic moved to Switzerland, where he had deposited $2 million in a Swiss bank account. Ivy, shunned by his colleagues, died in 1978, still convinced that Krebiozen was a cure for cancer.

By the 1980s, Max Gerson invented a miracle cancer diet. Gerson was a German-born physician who immigrated to the United States in 1936. Taking advantage of American's fear of environmental toxins, he offered a two-step cure: eat natural foods, consisting of a gallon of blended fruits, vegetables, and raw calf's liver, and remove harmful toxins with daily coffee enemas. Gerson's therapies also included liver extract injections, ozone enemas, "live cell therapy," thyroid tablets, royal jelly capsules, linseed oil, castor-oil enemas, clay packs, massive doses of vitamin C, and a vaccine made from killed staph bacteria. In the mid-1980s, three naturopaths—wanting to believe Gerson was onto something—visited his clinic, in Tijuana, and followed eighteen patients for five years; seventeen died of their cancers, and one still suffered the disease. The National Cancer Institute also reviewed the records of eighty-six patients treated with the Gerson diet and found no evidence that it worked—a finding that apparently didn't impress Steve Jobs when he chose Gerson's therapies instead of the surgery that might have saved his life.

The Gerson diet had other problems. Between 1980 and 1986, thirteen patients were admitted to local hospitals suffering from bloodstream infections caused by Gerson's liver extract injections. The Gerson Institute is still in operation.

I n the 1990s, it was shark cartilage.

On February 28, 1993, CBS's *60 Minutes* aired a story claiming that shark cartilage cured cancer. The segment starred I. William Lane, a Florida businessman who owned a shark-fishing company. Lane had written a book titled *Sharks Don't Get Cancer*. During the program, correspondent Mike Wallace showed several Cuban cancer victims exercising, all claiming that shark cartilage had cured them. Wallace was impressed, gushing about the new drug, cleverly called BeneFin.

Within two weeks of the *60 Minutes* episode, thirty new shark cartilage products were on the market. Within two years, shark cartilage was a $30-million-a-year industry. In 1996, Lane wrote *Sharks Still Don't Get Cancer*. By the late 1990s, one in four cancer sufferers took shark cartilage.

In preparing his program, Mike Wallace had either ignored or was unaware of several facts. First, according to the Smithsonian Institution's Registry of Tumors in Lower Animals, sharks *do* get cancer, including cancer of the cartilage. Second, before the show, Lane had already enlisted a group of Belgian researchers to test his product. It didn't work. Third, even if shark cartilage did cure certain cancers, eating it didn't make much sense. Acids and enzymes in the stomach destroy proteins contained in cartilage. (That's why diabetics don't eat insulin; they inject it.) Within a few years of the *60 Minutes* episode, thirteen studies showed that shark cartilage didn't cure cancers of the breast, colon, lung, or prostate. In 2000, I. William Lane and his company were banned from making unsubstantiated claims that shark cartilage cures cancer.

The shark cartilage fad took its toll, contributing to the possible extinction of the spiny dogfish shark and the blue shark, both of which are now at risk. Sadly, some people used shark cartilage instead of conventional therapies that would have saved their lives, most notably a nine-year-old Canadian girl with a brain tumor. After the tumor had been removed, doctors recommended a course of radiation and chemotherapy that offered a good chance of survival. Her parents opted for shark cartilage instead. She was dead in four months.

Shark cartilage is still available in drugstores and on the Internet.

Sick Children, Desperate Parents:

Stanislaw Burzynski's Urine Cure

You don't really understand human nature unless you know why a child on a merry-go-round will wave at his parents every time around—and why his parents will always wave back.

—William D. Tammeus, columnist, *The Kansas City Star*

Although many nonstandard therapies are available, the story of "alternative" cancer cures at the beginning of the twenty-first century has largely been the story of one man. A man whose early career was promoted by author Gary Null in *Penthouse* magazine, by Sally Jessy Raphael on her daily talk show, by Geraldo Rivera on ABC's *20/20*, and by Harry Smith on *CBS This Morning*. A man who, when hounded by the FDA, had the unsolicited support of hundreds of patients and several

United States congressmen. A man who, in 2010, was the subject of a feature-length documentary simply titled *Burzynski*.

B illie Bainbridge was born on April 25, 2007, in the town of Exeter, England. When she was four years old, her mother, Terri, noticed that something was wrong. "We had a cuddle in bed while she had a nap and I thought she seemed really hot," recalled Terri. "When she woke up, her head shook in like a mini-seizure and then she threw up." The next day, Billie was fine. Then she worsened. "Over the next couple of weeks she started talking in a weird voice and I thought it was just her messing around and kept telling her to stop. But then a few days later her eyes seemed a bit droopy and she was starting to [drool]. I thought she was having a stroke."

Two days later, Billie was taken to the Royal Devon and Exeter Hospital for an MRI of her head. The diagnosis: brainstem glioma, an incurable form of brain cancer. Brainstem gliomas are treated with radiation, which initially shrinks the tumor and improves symptoms. But the glioma invariably returns, eventually becoming resistant to radiation. Chemotherapy doesn't work, either. The doctors explained that Billie's chance of recovery was slim. In all likelihood, she would be dead in a year.

But Sam and Terri Bainbridge weren't giving up. They had heard about a miraculous medicine in Houston, and they were going to do to whatever it took to get it. The cost, however, was astronomical, estimated at about £200,000, or $300,000. And Terri was having her own problems, having recently been diagnosed with breast cancer. The Bainbridges took their case to their friends and neighbors. "I refuse to believe that Billie will die," said Terri. "And I'm going to do everything I can to make sure she doesn't."

Within weeks, everyone in England knew about Billie and Terri Bainbridge. Peter Kay, a British comedian and actor, staged a series of benefit concerts at the Blackpool Opera House. The band Radiohead donated a signed guitar, which sold for £9,000. Other bands, like Badly Drawn Boy, I Am Kloot, and Everything Everything, staged charity shows in Manchester. The Munchkins Day Nursery sold raffle tickets for £100, with a chance to win one year's tuition, worth £20,000. (The winners, Christian and Alex Brook, donated the money to the Bainbridges and continued to pay their son's tuition.) The local rugby team, the Exeter Chiefs, grew their sideburns, hoping to raise £2,000. A charity ball raised £12,000. Two cyclists raised more than £5,000 biking from Edinburgh to Exeter. A series of bake sales each raised £1,000. International celebrities like Anthony Cotton, Cheryl Cole, and Michael Bublé also pitched in. An anonymous philanthropist from the United States donated $25,000. In the end, the Bainbridges had raised £200,000. On September 17, 2011, only three months after Billie's first symptoms, the Bainbridge family was on a plane to Houston. "The Texas Clinic is our last hope," said Sam. Soon they would meet its director.

S tanislaw Burzynski was born in Poland in 1943. By the time he had graduated from the Medical Academy of Lublin, in 1967, he had published fourteen papers, a remarkable accomplishment. The following year he received a PhD for his thesis, titled "Investigations on Amino Acids and Peptides in Blood Serum of Healthy People and Patients with Chronic Renal Insufficiency." Burzynski had an idea. He believed that patients with long-standing kidney disease didn't get cancer as often as people with normal kidneys. And he believed the answer could be found in their urine. Burzynski reasoned that patients with kidney disease didn't get cancer because, unlike people with normal kidneys, they didn't excrete certain lifesaving substances that had an anti-cancer effect; he called these substances antineoplastons (literally, "against new growth"). Burzynski believed that if he could isolate these cancer-preventing peptides from the urine of healthy people, he could cure cancer.

After completing a residency in internal medicine in Poland, Burzynski traveled to Baylor Medical College, in Houston, where he rose from research associate in the department of anesthesiology to assistant professor. In 1974, he isolated a series of peptides (strands of amino acids smaller than proteins) that inhibited the growth of bone cancer cells in the laboratory. The National Cancer Institute, intrigued by his findings, gave him a three-year grant, which resulted in six publications—the last of which defined his life's calling. "According to our definition," he wrote, "antineoplastons are substances provided by the living organism that protect it against development of neoplastic growth [without] significantly inhibiting the growth of normal tissues." This was a major breakthrough. At last, cancer victims would no longer suffer the tortures of radiation and chemotherapy. They could be treated with antineoplastons, a natural product without side effects. The National Cancer Institute didn't see it that way, failing to immediately renew Burzynski's grant.

When he lost his funding, Baylor administrators gave Burzynski a choice. He could either stay in the department of anesthesiology and do research consistent with its goals (which weren't curing cancer) or he could leave. Burzynski left, renting a 2,500-square-foot garage in Houston that later became the Burzynski Research Institute. His first task was to isolate large quantities of antineoplastons. So in the heat of the Houston summer, Stanislaw Burzynski collected more than a hundred gallons of urine from public restrooms and local prisoners. From this he isolated antineoplastons that he called AS2.1 and A10.

Although Burzynski had no specific training in cancer, the FDA granted him permission to test antineoplastons in small clinical trials. He had no shortage of patients, including those with advanced cancers of the brain, colon, pancreas, breast, prostate, rectum, lung, kidney, and bladder—cancers so far gone, the medical establishment had given up on them, cancers that were a death sentence for their sufferers. But Stanislaw Burzynski wasn't giving up. He had faith in the power of antineoplastons.

In 1979, Gary Null, a popular critic of mainstream medicine, wrote an article in *Penthouse* titled "The Suppression of Cancer Cures." Null told the story of a sixty-three-year-old man with lung cancer that had metastasized to his brain. After six weeks of antineoplastons, the lung cancer was gone. After a few more weeks, the brain metastases were gone, too. According to the article, forty-one patients with advanced cancers had been treated, with "definite improvement in 86 percent." This was Burzynski's first major exposure to the media, and it dramatically increased his business.

Two years later, Geraldo Rivera, a correspondent for ABC's *20/20*, aired a show titled "The War on Cancer: Cure, Profit or Politics?" Again, Burzynski was cast as an unrecognized hero. Rivera compelled his audience with stories of hope and survival:

- "In early 1979, Steve Hipp got the dreaded news," said Rivera. "Steve had cancer. His doctor in Michigan summed it up this way: 'We were waiting for him to exit.' Steve's doctor recently confirmed to *20/20* that his tumor has shrunk."

- "Al Swazeland is from Canada," said Rivera. "In late 1978, Al discovered that he had bladder cancer. Six operations later, his only chance was removal of his bladder, which meant he would need a urine bag in his abdomen for the rest of his life. Al just informed us that as of now he has no active tumors."

- "Jocelyn Chancey's inoperable breast cancer had spread to her bones," said Rivera. "I talked to about fifteen doctors and the consensus was that she was terminal," said her husband. But like Steve Hipp and Al Swazeland, Jocelyn Chancey had had a miraculous response to antineoplastons. "Jocelyn Chambers's bone scan shows that she is in partial remission," said Rivera, "and a more recent bone scan shows further improvement." "I sleep good at night," said Jocelyn.

Only four years after Stanislaw Burzynski had opened the Burzynski Research Institute, he was curing patients with no hope of a cure. Still, antineoplastons hadn't entered the mainstream. Rivera knew why. "There is a cancer establishment," he said, "and it's basically divided into two parts. One is our most lavishly funded government health agency. It's called the National Cancer Institute. The other is our wealthiest private charity, the American Cancer Society. To critics, their combined function has a stranglehold effect, creating a kind of monopoly on cancer research and information." According to Rivera, cancer doctors weren't sympathetic caretakers; they were ruthless businessmen. "Cancer is not just a disease," he said; "it's a political and economic phenomenon, a $30-billion-a-year business"—one that apparently had no intention of including the likes of Stanislaw Burzynski. "If somebody were able to bring some new, innovative idea into the fight against cancer," Burzynski told Rivera, "then finally the American public will ask the big institutions, 'What are you doing with all this money? Where are your results?' Finally, they would have to answer to the American public!"

The Burzynski miracle marched on. Seven-year-old Dustin Kunnari received antineoplastons for a brain tumor; six weeks later, the tumor was gone. Tori Moreno's advanced brain cancer was gone in five months. Antineoplastons had also cured the brain tumors of Pamela Winningham, Crystin Schiff, Zachary McConnell, and Thomas Wellborn. And they had cured Randy Goss's kidney cancer.

In 1995, Harry Smith, of *CBS This Morning*, brought several patients cured by antineoplastons onto his show. "It's like cancer never happened to me," said Neal Dublinsky, of Los Angeles. "Houston was the city where my life was saved." In 2001, Thomas Elias wrote *The Burzynski Breakthrough: The Most Promising Cancer Treatment and the Government's Cam-*

paign to Squelch It. Elias told the story of Burzynski's persecution by the medical establishment. And he told the stories of patients who had been saved by antineoplastons despite the best efforts of cancer specialists to ignore the miracle in front of them.

In 2010, the documentary *Burzynski* aired throughout the country. Written and directed by Eric Merola, *Burzynski* featured the recovery of Jodi Fenton and Jessica Ressel from brain cancer and Kelsey Hall from adrenal cancer—all because of antineoplastons, all despite the grim prognoses of so-called cancer specialists. The movie also showed attempts by FDA commissioner David Kessler to shut down Stanislaw Burzynski. "It constitutes nothing less than one of the worst abuses of the criminal justice system I have ever witnessed," said Congressman Richard Burr of North Carolina. Burzynski sat calmly in the midst of the maelstrom, unwavering, undaunted. He was a modern-day Semmelweis whose demonstration that hand washing could stop the spread of germs was ignored by his peers. "It is obvious that Dr. Burzynski has made the most important discovery in cancer ever," said Dr. Julian Whitaker.

By 2011, the Burzynski Research Institute, located in Stafford, Texas, housed four chemists, four biologists, three pharmacists, and an antineoplaston production facility occupying more than 46,000 square feet—the size of a small biotech company.

S o why haven't antineoplastons become part of routine cancer care? Surely patients haven't made up their stories—surely people don't say they've been cured of terminal cancer when they haven't or say they're still alive when they're not. A closer look, however, shows that Stanislaw Burzynski's antineoplaston cure isn't exactly what it's claimed to be.

In November 1982, several Canadian residents who had received antineoplastons wanted to be reimbursed by their insurance company. In response, Canadian researchers Martin Blackstein and Daniel Bergsagel traveled to Houston. This was the first independent review of Burzynski's program. Blackstein and Bergsagel asked Burzynski to provide four pieces of information: (1) a biopsy proving that the patient had had cancer; (2) records of treatments before antineoplastons; (3) records of treatments with antineoplastons; and (4) radiological studies like X-rays and CT scans. Burzynski provided data from his fourteen most successful cases.

Although Blackstein and Bergsagel had been clear about what they'd wanted, what they'd gotten was disappointing. Records were scant and incomplete. Two patients had CT scans before and after antineoplastons: no change. Two other patients had been cured by standard treatments *before* seeing Burzynski. Blackstein and Bergsagel couldn't find any objective evidence that antineoplastons worked. "We were surprised that Burzynski would show us such questionable cases," they wrote. "We were left with the impression that either he knows very little about cancer . . . or else he thinks that we are very stupid and has tried to hoodwink us."

Blackstein and Bergsagel also asked to see the records of four other patients whom Burzynski had described in a medical journal in 1977. Again, the results were disappointing. Three of the patients had died from their cancers. The fourth had been cured by surgical removal of a bladder tumor.

The second independent review of Burzynski came in 1985, again from Canada. This time investigators from the Canadian Bureau of Prescription Drugs reviewed the records of thirty-six patients. Thirty-two had died from cancer without benefit from antineoplastons. Of the four remaining patients, one had died after a slight regression of the tumor; one had died after being stable for a year; and two were still alive at the time of the survey. Of the two who were still alive, one had metastatic lung cancer and the other cervical cancer—neither had been cured.

In 1988, Sally Jessy Raphael, a popular talk show host, interviewed four patients treated by Burzynski whom she described as "miracles." Four years later, *Inside Edition* followed up. Two of the four had died of cancer, and a third was in the midst of a recurrence. The fourth had been cured of a bladder cancer that had had an excellent prognosis.

In 1990, the United States Congressional Office of Technology Assessment reviewed journal articles published by Burzynski and concluded, "Despite a substantial number of preliminary studies, there is still a lack of valid information to judge whether this treatment is likely to be beneficial to cancer patients."

As patient testimonials mounted, independent researchers offered to test antineoplastons. One was Joseph Jacobs, director of the newly created Office of Alternative Medicine

(OAM), later to become the National Center for Complementary and Alternative Medicine. Burzynski resisted. "OAM was willing to buy the research assistance for [Burzynski] to design a good protocol and to set up a data-monitoring committee," said Jacobs. "There have been plenty of opportunities. And those clowns—his supporters—were doing everything they could to wreck those opportunities."

In 1991, two investigators from the National Cancer Institute visited the Burzynski Research Institute and, after reviewing clinical histories, pathology slides, and radiological studies from seven patients, concluded that he might be onto something. They agreed to fund an expensive clinical trial in patients with brain tumors. This was Burzynski's single best chance to convince the FDA to license his drug.

After years of haggling with Burzynski about which patients should be included, the National Cancer Institute published its study. The collaboration involved researchers from the Mayo Clinic in Rochester, Minnesota, the Memorial Sloan-Kettering Cancer Center in New York City, and the National Cancer Institute, in Bethesda, Maryland—three of the most prominent cancer centers in the world. It was the first time a meaningful publication about antineoplastons didn't have Stanislaw Burzynski's name on it. The results were disappointing. All nine patients treated with antineoplastons died; none had had any evidence of a response to treatment. And despite Burzynski's claims to the contrary, antineoplastons were toxic: some patients suffered nausea, vomiting, headaches, and muscle pain; others, excessive sleepiness, confusion, and seizures. Burzynski was furious. Convinced that the National Cancer Institute had purposefully ruined his work, he said, "They sabotaged the trial. They were trying to give the treatment for a very short period of time, like for instance a couple of weeks or months. And then, of course, the patient was dying after that. It was completely unethical. It was horrible."

In 1992, Saul Green, a biochemist, summarized what was known about antineoplastons AS2.1 and A10 in a paper published in the *Journal of the American Medical Association*. He explained that AS2.1 was phenylacetic acid (PA), a potentially toxic substance produced during normal metabolism. Typically, phenylacetic acid is detoxified by the liver—where it becomes phenylacetylglutamine (PAG)—and is then excreted in the urine. A10 is essentially PAG. Burzynski had claimed that AS2.1 and A10 worked by inserting themselves into DNA and changing the genes that cause cancer. However, PA and PAG are too big to fit into DNA. Green concluded, "The treatment for cancer with substances called antineoplastons actually involves the use of two simple commercially available organic chemical compounds, PA and PAG. None is a peptide, none has been shown to 'normalize' tumor cells, none has been shown to actually [insert] into DNA, and none has been shown to be active against cancer in experimental tumor test systems."

Further investigations led to further disappointments.

During the 1990s, three independent reviewers evaluated 963 patients who had received antineoplastons. All three were cancer specialists: Howard Ozer, director of Allegheny University Cancer Center in Pittsburgh; Henry Friedman, professor of pediatrics at Duke University and chairman of the brain tumor committee of the Pediatric Oncology Group; and Peter Eisenberg, a community oncologist practicing in Marin County, California. All agreed that Burzynski's protocols were poorly designed, his data uninterpretable, and the toxicities of antineoplastons potentially dangerous. Most upsetting was Burzynski's treatment of a four-year-old boy with a potentially curable brain tumor. Treated with antineoplastons only, the boy's cancer worsened.

By 1995, the FDA had had enough, charging Stanislaw Burzynski and his clinic with seventy-five counts, including criminal contempt, mail fraud, and violations of the Food, Drug, and Cosmetic Act. Congressman Joe Barton of Texas rushed to Burzynski's aid, holding a hearing that allowed others to tell their side of the story. Patients and parents carried placards proclaiming, "Say No to Chemo," while chanting, "FDA go away! Let me live another day!" In 1997, the government failed in two attempts at conviction: one trial ended in a hung jury, the other in a not-guilty verdict.

For many cancer sufferers, Stanislaw Burzynski had become a countercultural hero. But not to everyone. "It's a travesty of everything we fought for as activists," said Fran Visco, president of the National Breast Cancer Coalition. "If this is the type of research that is permitted to go forward, it's a threat to our lives and a threat to continued support of science. How is he getting away with it?" Ellen Stovall, executive director of the National Coalition for Cancer Survivorship, agreed: "From this moment on we are not going to let him rest. He is insulting the intelligence of the American people by calling his therapy nontoxic and alternative. I would like to see Dr. Burzynski's congressional patrons apologize to the American people."

During the 1980s and '90s, three companies showed interest in antineoplastons: Sigma Tau Pharmaceuticals of Italy, Chugai Pharmaceuticals of Japan, and Elan Pharmaceuticals of Ireland. One by one they dropped away. The medical director of Sigma Tau wrote, "Dr. Burzynski was informed on January 31, 1991, that Sigma Tau did not intend to proceed with the development of the antineoplastons. We have studied antineoplastons A10 and AS2.1 . . . in human and [mouse] tumor cell lines. . . . On the basis of these results the project has been discontinued." Burzynski saw the rejections as a conspiracy among cancer doctors willing to lie to save their careers. "Most of the oncologists—I'm talking about reputable oncologists—they work for pharmaceutical companies," said Burzynski. "They work in clinical trials; they receive various types of incentives from pharmaceutical companies. They will do everything they can to lie. . . . We have a lot of evidence about some of these doctors who are dishonest, who are liars, who cheat."

In truth, it's not that hard to prove that new cancer therapies work. Paul Goldberg, editor of *The Cancer Letter*—a Washington, D.C.–based publication covering cancer research and drug development—has commented on the strange case of Stanislaw

Burzynski. "Drug approval is a technical, time-consuming, and costly process," he said. "Nonetheless, thousands of anticancer compounds have been shown to be effective—or dismissed as ineffective—over the decades since Dr. Burzynski initiated his experiments." Henry Friedman, the oncologist from Duke University who had independently reviewed Burzynski's data, agreed: "Despite thousands of patients treated with antineoplastons, no one has yet shown in a convincing fashion . . . that the therapy works. You have to understand how incredible that is. Because normally you can do a cancer study on as few as thirty or forty patients."

One cancer drug's story is particularly instructive. In 2002, an article was published in the scientific journal *Nature* showing that malignant melanoma cells contained an abnormal gene called *BRAF*. Investigators reasoned that a drug that blocked the protein made by this gene could work. In response, a small biotech company in Berkeley, California, called Plexxikon— not much bigger than the Burzynski Research Institute—took an interest. They made a drug called PLX4032 and—under the auspices of Paul Chapman, an oncologist at Memorial Sloan-Kettering Cancer Center—tested it on thirty-two patients with advanced melanoma. All saw their tumors shrink. The standard chemotherapy at the time, dacarbazine, slowed tumor growth in 15 percent of patients for two months. PLX4032, on the other hand, halted tumor growth in 80 percent for eight months. Encouraged, company researchers performed a definitive trial. They randomly assigned 680 patients to receive either dacarbazine or PLX4032. The results were clear: patients who received PLX4032 lived longer. On August 17, 2011, the FDA approved PLX4032 for the treatment of malignant melanoma.

The process took only a few years and involved fewer than a thousand patients.

Stanislaw Burzynski has administered antineoplastons to thousands of patients for decades and has never published a definitive study comparing his drug with standard therapy. "He's convinced there's no need for studies," says Peter Adamson, chairman of the Children's Oncology Group at the Children's Hospital of Philadelphia and professor of pediatrics and pharmacology at the Perelman School of Medicine at the University of Pennsylvania. "He runs a risk with no benefit. Everyone who comes there is paying him to do it, so he has absolutely nothing to gain and everything to lose."

I n the world of alternative cancer cures, Stanislaw Burzynski is a strange case. Not only is he a brilliant man—one of the youngest to hold both MD and PhD degrees from his native Poland—he also presents himself as a caring man. Craig Malisow, of the *Houston Post*, wrote, "Patient testimonials almost always contain the phrase 'treated like family.' His warm bedside manner, and that of his staff, is often in stark contrast to patients' experiences in mainstream hospitals." For these reasons, authors like Thomas Elias and filmmakers like Eric Merola and celebrities like Sally Jessy Raphael and Geraldo Rivera and reporters like Harry Smith paint Burzynski as an unappreciated hero. But too many things about Stanislaw Burzynski don't add up. He claims to have a cure for cancer, but when the Office of Alternative Medicine offered to test it, he balked. And when researchers at the National Cancer Institute gave him a chance to bring his ideas into the mainstream, he argued they didn't know what they were doing. As a result, four decades have passed and we still don't have any convincing scientific evidence that antineoplastons work.

S till, one could argue, where's the harm? Burzynski rarely takes care of patients with treatable cancers. He takes care of patients, like Billie Bainbridge, whose prognoses are grim. "He's not treating children with leukemia or Wilms tumor," admits Peter Adamson. But false hope, he argues, isn't a gift: "When I meet with families whose cancer has no curative path, the discussion is 'How can we leave as many options open as possible?' When an experimental option shows some promise, we move it up the list. But when an experimental option—like Burzynski's—is ineffective, we take it off the list. It doesn't take that many children to know that something doesn't work."

John Maris, chief of the division of oncology at the Children's Hospital of Philadelphia, agrees. "If you're diagnosed with brainstem glioma [like Billie Bainbridge]," says Maris, "your chance of cure is very low. But there are therapies that can extend life and extend life with quality. What many of us in the area of refractory cancers are trying to do is take a page out of the AIDS paradigm—try to make it a chronic disease and hope that science can catch up and offer new realistic therapies. I worry when patients who could benefit from evidence-based investigational therapies bail out and go to Houston [to see Burzynski]. I am quite certain as a scientist that antineoplastons are completely worthless. And I say that as someone who has spent a fair amount of time looking for new therapies for cancer, because I'm sick of watching children die from this disease."

In September 2011, Billie Bainbridge began antineoplaston therapy. During the first five weeks of increasingly greater doses, she got sicker and sicker. She lost her appetite and started vomiting. At one point she had to be admitted to Texas Children's Hospital. Facing severe dehydration, malnutrition, and weight loss, Billie had to have a feeding tube placed into her stomach. Because the Burzynski Clinic doesn't have a clinical facility, and because Burzynski doesn't have admitting privileges to Texas Children's Hospital, her care was out of his hands.

Perhaps no one has a better view of Stanislaw Burzynski's operation than Jeanine Graf, an associate professor at Baylor College of Medicine and medical director of the pediatric intensive care unit at Texas Children's Hospital. Graf has taken care of several of Burzynski's patients at the end of their lives. "Why do this?" asks Graf, arguing that Burzynski's treatments only lead to greater tragedy. "It only removes families from their home, putting incredible stress on them. I'm forced to say that 'we don't have anything to offer, and now your child is in an ICU, and do you want to have your child die in an unfamiliar place with people who don't know you, no family around?' That is just not the right way to end your life."

Graf occasionally faces the task of figuring out how to

get critically ill children whose parents are now broke back to their homes, surrounded by family and friends. "We have picked up the tab to transport some of these children back home because they've come from so far and they're absolutely depleted of their finances," says Graf. "Luckily, we have a robust charity committee. I can make a case that this is a compassionate thing to do for this family and we do it."

In October 2011, Billie returned to England, continuing to receive antineoplastons intravenously—and continuing to suffer from the drug. "It is making her more and more tired," said Terri, "which is making it difficult to get her to eat enough." By late October the vomiting had again become so severe that her treatment was stopped and she was admitted to a local hospital with dehydration.

In England, where Billie Bainbridge continued to receive national attention, child advocates and bloggers commented on the case. One science blogger summed it up best. Referring to the £200,000 raised by the Bainbridges to pay Stanislaw Burzynski, he wrote, "Peter Kay is right to raise money for this family. And good luck to him. But it would be a dreadful wrong for this money to end up in the hands of someone [else]. The money could make a big difference to this family. It could allow both mother and daughter to be looked after in comfort, without worrying about mortgages or jobs. It will allow them to be together. It will not perform miracles. Nor will it make the pain go away. But such a simple gift will indeed be an act against cynicism and false hope."

On June 1, 2012, one year after her diagnosis, Billie Bainbridge lost her battle with cancer. "She was incredibly brave," said her family, "and never complained or asked why."

Burzynski's most recent activities have been even more demoralizing. He now claims that antineoplastons treat not only cancer but Parkinson's disease, AIDS, and neurofibromatosis. Burzynski has also taken a step that puts him firmly in the cabinet of sideshow hucksters who have been selling their cure-alls since the first traveling carnival. He is promoting a line of creams and capsules with the brand name Aminocare, described as "The Genetic Solution for Anti-Aging." For $120, you can buy Aminocare A10 gel caps, which "aids the regulation of normal cell division"; for $50, Aminocare Skin Cream, which "slows the signs of aging by stimulating natural cell division"; and for $60, Aminocare Brain Longevity Forte, which "helps maintain normal brain function."

Perhaps no one has offered a bigger stage for Stanislaw Burzynski's antineoplastons and other alternative cancer cures than Suzanne Somers, who is promoting them in her books as well as on CNN, MSNBC, and Fox. In *Knockout: Interviews with Doctors Who Are Curing Cancer and How to Prevent Getting It in the First Place*, Somers touts antineoplastons, coffee enemas, and miracle diets. The book stands in stark contrast to another book published at the same time, Siddhartha Mukherjee's *The Emperor of All Maladies: A Biography of Cancer*. Both tell the story of cancer. And both have sold well. Only one,

Mukherjee's, won the Pulitzer Prize. And only one, Mukherjee's, tells the story of cancer from a scientist's perspective.

Mukherjee pulls no punches. From ancient times to the modern era, the story of cancer treatment has been one of collateral damage. Surgery removes cancerous tissue as well as normal tissue. Radiation and chemotherapy kill cancer cells as well as normal cells. But Mukherjee's book also describes a new trend in cancer therapy: specificity.

During the past few decades, scientists have begun to identify cancer-causing genes called oncogenes. And they've developed drugs like Herceptin and Gleevec that target the products of these genes, a major advance. Because these drugs are specific for cancer cells, their side effects are far more tolerable than those from standard chemotherapy. Gleevec, for example, has changed the face of one type of chronic leukemia in adults, a diagnosis that at one time was a death sentence. Now patients can survive for decades.

Nowhere in Somers's book do we learn about oncogenes and their products, and nowhere in Mukherjee's do we learn about coffee enemas and miracle diets. It's as if they were written in parallel universes. In Mukherjee's universe, drugs have to be science-based, thoroughly tested, and proven to work before they're licensed by the FDA. In Somers's universe, treatments aren't science-based, proven to work, or licensed by the FDA; rather, they're promoted with testimonials and sold on websites.

What is perhaps most disappointing is that television producers have consistently chosen Somers over Mukherjee to educate their viewers. Siddhartha Mukherjee is a Rhodes Scholar, an assistant professor of medicine at Columbia University Medical Center, and a graduate of Stanford University, the University of Oxford, and Harvard Medical School. He has devoted his life to caring for cancer victims and researching ways to cure them. Suzanne Somers was Chrissy on the hit TV series *Three's Company* and the promoter of the popular Thigh-Master. She has spent much of her life extolling unproven cures in books and selling them on her website. To producers, the choice of Somers over Mukherjee has apparently been an easy one.

Part VI

CHARISMATIC HEALERS ARE HARD TO RESIST

10

Magic Potions in the Twenty-First Century:

Rashid Buttar and the Lure of Personality

Me and Shrek took some magic potion, and now
we're sexy!
—The Donkey, *Shrek 2*

My father was the head of a sales force that sold men's shirts.
Every six months, salesmen from across the country would
meet in Baltimore and my father would teach them how to sell.
His message was clear: salesmen weren't selling shirts; they
were selling themselves.

Even though I was just a little boy, my father let me go to
those meetings. (I went for the food.) But I still remember the
names of most of those salesmen. I really loved those guys.
They were funny and affable and kind. And even though I

knew that their stories were exaggerated or fabricated, it didn't matter. I enjoyed being around them.

My other childhood memory of the lovable huckster came in the form of a *Twilight Zone* episode that aired in October of 1959. Titled "One for the Angels," it featured two veteran actors: Ed Wynn, the rubber-faced comic who starred as the Toymaker in *Babes in Toyland* and Mr. Dussell in *The Diary of Anne Frank*, and Murray Hamilton, best known as the mayor of Amity in the 1975 movie *Jaws*.

The episode opens with Wynn standing behind a suitcase propped on a wooden stand. "Right here, ladies and gentlemen," he shouts. "Special July clean-up sale!" Rod Serling, host of *The Twilight Zone*, sets the scene: "Man on the sidewalk named Lou Bookman. Age: sixtyish. Occupation: pitchman."

Bookman slowly closes his suitcase, folds up the stand, and returns to his stoop, where he is immediately mobbed by several children. "What are you selling today, Lou? Toys?" asks one, an eight-year-old named Maggie. Bookman, who loves children, gives a wind-up robot to each of them.

When Bookman returns to his apartment, he finds Murray Hamilton sitting on a chair, thumbing through a small notepad. Hamilton is the angel of death. After confirming Bookman's age, birthplace, employment history, and parents' names, he says, "Your departure is at midnight." "My departure?" asks Bookman, horrified. To remove any doubt about his intentions, the angel touches a flower, which wilts and dies.

Bookman pleads for his life. "But I have some unfinished business," he says. "Between you and me, I've never made a truly big pitch, enough for the skies to open up. You know, a pitch for the angels." The angel relents, allowing Bookman one last pitch. But Bookman has no intention of keeping his promise. When the angel realizes that Bookman has conned him, he selects an alternate.

On the street in front of Bookman's apartment, the sound of screeching tires is followed by screams. Maggie, unconscious, is lying on the ground. Seeing the consequences of his actions, Bookman recants. "Take me!" he pleads. "She's just a little girl. She's only eight years old." But it's too late. The angel of death tells Bookman he'll be back at midnight to claim her.

Fifteen minutes before midnight, the angel arrives at the stoop of Maggie's apartment. Bookman, who has been waiting for him, opens his suitcase and takes out an ugly cotton tie. "Take this lovely tie here, for instance," he says. "What does this look like to you?" "It looks like a tie," the angel deadpans. "Ladies and gentlemen," expounds Bookman. "If you will feast your eyes on probably the most exciting invention since atomic energy. A simulated silk so fabulously conceived as to mystify even the ancient Chinese silk manufacturers—an almost unbelievable attention to detail. A piquant intervening of gossamer softness." Riveted, the angel buys the tie.

Next, Bookman holds up an ordinary spool of thread. "This fantastic thread is not available in stores," he says. "It is smuggled in by Oriental birds specially trained for ocean travel, each carrying a tiny thread in a small satchel underneath their ruby throats. It takes 832 crossings to supply enough thread to go around one spool." The angel can't reach for his wallet fast enough. "I'll take all you have," he says.

Bookman continues: "Sewing needles, marvelous plastic

shoelaces, genuine static eradicator, suntan oil, eczema powder, razors, athlete's foot destroyer. How about some simulated cashmere socks?" The angel is sweating, obsessed: "All right, all right. I'll take it all!"

The clock strikes midnight—too late for the angel to claim Maggie. The angel realizes he's been had. "One minute past twelve, Mr. Bookman," he says. "And you made me miss my appointment." "Yes," replies Bookman. "It was quite a pitch—very effective. The best I've ever done. A pitch so big, the sky would open up."

> Serling in voiceover: "Louis J. Bookman. Age: sixtyish. Occupation: pitchman. Formerly a fixture of the summer. Formerly a rather minor component to a hot July. But throughout his life, a man beloved by children. And, therefore, a most important man. Couldn't happen, you say? Probably not in most places. But it did happen—in the Twilight Zone."

When Louis Bookman was at the height of his sales pitch—knowing he had to distract the angel of death if Maggie were to survive—he talked in a rapid-fire, high-pitched, nasal voice, like a duck. In the sixteenth century, the Dutch had a name for this: *kwakzalver*, meaning one who quacks like a duck while promoting salves and ointments. This became the English word *quacksalver*, later shortened to *quack*, meaning anyone who proffers false cures. For some, the term also implies intent—that quacks are knowingly fraudulent in their pursuit of fortune. But this isn't always the case.

In many ways, Bookman was a classic quack. He was an impassioned salesman who clearly loved children and wanted to protect them. He was convincing, in large part, because he was convinced. When Bookman claimed that 832 Oriental birds were required to make one spool of thread, he believed—at least for the moment—in what he was saying, even though it was pure fantasy. And his cures for eczema and athlete's foot were offered well before treatments like topical steroids and antifungal creams were widely available. Bookman's false promises weren't unusual. In the 1800s, quacks offered Dill's Diabetic Mixture before the discovery of insulin; Peebles' Epilepsy Treatment before anti-seizure drugs; Dr. Shoop's Diphtheria Remedy before diphtheria antiserum; Detchon's Rheumatism Cure before anti-inflammatory drugs; Az-Ma-Syde before bronchodilators; William Radam's Microbe Killer before antibiotics; and Cancerine before chemotherapy.

Quacks offered potions that made you smarter (Harper's Brain Food), younger (Blush of Youth), less anxious (Dr. Kline's Great Nerve Restorer), more successful (Wendell's Ambition Pills), less freckled (Dr. Berry's Freckle Ointment), more potent (Las-I-Go for Superb Manhood), and more fertile (Becket's Sovereign Restorative Drops for Barrenness). Patent medicines had such wonderful, colorful names, you couldn't help but buy them: medicines like Admirable Essence of Life, Squire's Grand Elixir, Hamlin's Wizard Oil, and Kickapoo Indian Sagwa (satirized in the cartoon *Li'l Abner* as Kickapoo Joy Juice). And it wasn't only the public that believed these claims; celebrities did, too. Devices like the Radio X pad made Al Jolson a better singer, and "nuxated" iron made Jack Dempsey a better fighter and Ty Cobb a better hitter. At least according to them.

We look back warmly on these salesmen and their funny medicines—a whimsical, bygone era made obsolete by the relentless advances of science. But there's no need for nostalgia—hucksters and their wondrous elixirs haven't gone anywhere. One works in a small town just outside Charlotte, North Carolina. In fact, people travel from all over the world to see him—and to buy the two magical potions he invented.

R ashid Buttar graduated from Washington University in St. Louis, majoring in biology and theology before attending the College of Osteopathic Medicine and Surgery, in Des Moines, Iowa, where he specialized in emergency medicine. He's enormously popular, attracting patients from thirty-six states and forty-two countries.

Buttar has also written a book—*The 9 Steps to Keep the Doctor Away*—and produced a series of instructional videos such as *Heavy Metal Toxicity: The Hidden Killer*; *Autism: The Misdiagnosis of Our Future Generations*; and *Cancer: The Untold Truth*. Both as a writer and as a speaker, Buttar delivers his message passionately, clearly, and compellingly. He has been quoted in the *Wall Street Journal*, *US News & World Report*, and the *New York Times* and appeared on ABC's *20/20*, PBS's *Frontline*, and CBS's *World News Roundup*. In May 2004, Buttar testified before a congressional committee investigating new treatments for autism.

Buttar's message is simple: environmental toxins such as mercury and lead cause chronic illnesses, which should be treated with chelation medicines. (Chelation, from the Greek *chele*, meaning "claw," binds heavy metals and rids them from the body.) Rashid Buttar knows what scares people. From Rachel Carson's 1962 book *Silent Spring*—warning of the dangers of DDT—to concerns about environmental toxins today, it's easy to appeal to the notion that we're poisoning ourselves. And the need to keep the body pure is centuries old, reflected in the text of every major religion. In *9 Steps*, Buttar writes, "If it's in the form God created, it's good. If it's not, leave it alone. God given = Good. Man-made = Madness." To rid themselves of unseen toxins, hundreds of thousands of Americans receive chelation drugs every year, usually intravenously.

Although popular, the fear that man-made poisons are causing chronic illnesses is largely unfounded. Studies haven't supported the concern that certain environmental contaminants such as dioxin, radon, bisphenol A, hexavalent chromium (the villain in *Erin Brockovich*), trichloroethylene (the basis of the book and movie *A Civil Action*), and even DDT cause the diseases claimed. And while chelation therapy is valuable, it's not a panacea—it's required only for people exposed to large quantities of heavy metals, such as lead paint in old houses or methylmercury in contaminated fish.

R ashid Buttar doesn't see it that way. When patients come to him with cancer, he often chelates them. The same is true if they have arthritis, autism, diabetes, heart disease, Parkinson's disease, Lou Gehrig's disease, or hormonal problems. During a licensing hearing in April 2008, Dr. Art McCulloch, an anesthesiologist from Charlotte, asked Buttar's nurse practitioner, Jane Garcia, whether it seemed odd that every single one of their patients had been poisoned by heavy metals:

At the beginning of his career, Rashid Buttar didn't treat many children. Then something happened. "In January of 1999, my son, Abie, was born," said Buttar, fighting back tears. "At ten months old, he started to speak. He had a ten-, twelve-word vocabulary." But at fourteen months of age, Abie regressed and could no longer speak. The first word he'd lost was the first word he'd learned: *abu*, meaning "father" in Arabic.

Buttar soon realized that his son had autism and that God was asking him to do something about it. "Looking back, it's clear that God had a specific plan for me; but I was moving away from the right path," recalled Buttar. "My name in Arabic means 'one who stays on the right path of life.' Now I realize that this experience was nothing more than God upping the ante, sending me a clear message: 'You are going to do what you were meant to do, what you were created to do!'"

Buttar had seen his destiny. He would find a cure for autism. "I subsequently spent thousands of hours—many if not most of them late at night, sometimes all night—studying, researching, learning, crying, and praying that my son would be returned to me," wrote Buttar. "I pleaded, begged, and threatened God. I bartered with the Creator, negotiating my arms and legs in exchange for the return of my son."

Within a few years Buttar had developed a novel chelation therapy, one that didn't have to be injected or ingested, as all FDA-approved drugs for the treatment of true heavy-metal poisoning require. Rather, Buttar's chelation, called TD-DMPS (for transdermal dimercaptopropanesulfonic acid), could simply be rubbed onto the skin. The results, according to Buttar, were phenomenal. "Five months after I began his detoxification," wrote Buttar, "Abie went from no language to a vocabulary of five hundred–plus words. [Today] he's extraordinary—ahead of his peers in school in all subjects and two to three grade levels ahead in math and English, an incredible athlete in every sport he tries."

By April 2006 Buttar had treated more than 250 autistic children with his wondrous anti-autism cream. To determine whether his medicine was working, he tested children's urine, finding large quantities of mercury and lead. As toxins poured out, autistic children recovered, some dramatically. To many parents, Rashid Buttar was a hero, his drug a miracle. Unfortunately, Rashid Buttar, his therapies, and his diagnostic tests aren't quite as advertised. His inconsistencies came to light in a story that swept the nation and went viral on YouTube. It involved an NFL cheerleader with a terrible problem.

On August 23, 2009, a twenty-five-year-old cheerleading ambassador for the Washington Redskins named Desiree Jennings got a flu shot. Two weeks later, she developed a bizarre series of symptoms. She couldn't walk without flailing her arms and legs, and her speech was halting and robotic. Although Desiree couldn't walk, she could run, competing in an eight-kilometer race. She could also walk sideways and backwards. She just couldn't walk forward. Desiree claimed that if she listened to startling sounds like a telephone ringing or hip-hop or techno music, her symptoms worsened. However, if she listened to the English alternative rock band Coldplay, her symptoms improved. She also developed a British accent, even though she was born and raised in Ohio.

Desiree was taken to hospitals in Leesburg and Fairfax, Virginia, before ending up at Johns Hopkins Hospital, in Baltimore. There, she was examined by internists, physical therapists, speech therapists, neurologists, neuropsychologists, and psychiatrists who subjected her to a dizzying array of blood tests, scans, and metabolic screenings. Despite extensive testing, no one could find anything wrong with her. Finally, a physical therapist offhandedly provided Desiree with the name of a disease she could embrace: dystonia, a movement disorder.

On October 13, 2009, WTTG-5, a Fox network affiliate in Washington, D.C., picked up the story. Three days later, *Inside Edition* jumped in, opening its segment with "She's the beautiful cheerleader whose heartbreaking story is shocking the nation!" Footage from *Inside Edition* made it to YouTube. Soon hundreds of thousands of people had learned that the flu vaccine caused a horrible, disabling disease.

Anti-vaccine activists rushed to Desiree's aid. Jenny McCarthy and then-boyfriend Jim Carrey, working with Generation Rescue—a group dedicated to the notion that a mercury-containing preservative in vaccines caused autism—directed

Desiree to the popular physician in North Carolina who they knew could cure her. In November 2009, Buttar examined Desiree at his clinic. His diagnosis was predictable: "mercury toxicity" from the flu shot. Buttar began intravenous chelation. "We took the toxins out of her system," declared Buttar, who confidently predicted that Desiree would fully recover. Within a few hours Desiree was feeling better. Amazing. So amazing that a film crew from ABC's *20/20* traveled to North Carolina to document what had happened. Unfortunately, with the cameras rolling, Desiree regressed. No longer able to walk, she had to be taken out of Buttar's clinic in a wheelchair.

Slowly, Desiree's story fell apart. First it came to light that neurologists at Johns Hopkins had diagnosed Desiree's problem as psychological. Later, other physicians weighed in. Yale neurologist Steven Novella wrote, "Jennings' movements [and] evolving speech patterns do not fit any known pattern of neurological damage. Rather, they are all features of psychogenic symptoms. The one that is probably the easiest for people to understand is her vaguely British accent. . . . There are only so many ways that speech can be neurologically abnormal—none of them make you sound British." Neurologists at the University of Maryland School of Medicine now use Desiree's YouTube video to illustrate what psychological movement disorders look like.

When the producers of *Inside Edition* realized they'd been had, they did a follow-up story. On February 5, 2010, they caught up with Desiree outside a shopping mall. "When Jennings first walked out of a store and into the shopping-center parking lot," said the correspondent, "she seemed to be walking normally. But as she left to get into her car [and saw our camera], she was walking sideways."

In the name of helping Desiree Jennings, Rashid Buttar had ignored the real cause of her problem. Desiree needed psychological support, not chelation. "I remain sympathetic to Desiree Jennings," wrote Novella. "She is an unfortunate woman who is being exploited by the media . . . and the anti-vaccine movement. What she needs is the delicate management of science-based practitioners who know how to deal with such cases." Later, Desiree said, "If I have to go over to China and do experimental procedures, I'll find a way to get [my life] back. It may take a while, but I will get everything back. I will find a way."

B uttar believes that his chelation medicines work on people like Desiree Jennings because he detects heavy metals in the urine after treatment. Unfortunately, Buttar's tests and conclusions are misleading, for several reasons.

First: Because metals like mercury and lead are present in the earth's crust, everyone has small quantities in their bloodstream. These trace quantities aren't harmful.

Second: Because everyone has small quantities of heavy metals in their bloodstream, virtually everyone who is given a chelating agent will excrete heavy metals in their urine.

Third: Reference ranges for heavy metals present in the urine after chelation don't exist. So when Buttar claims that patients have too many heavy metals in their body, he's groping in the dark. Indeed, a look at the fine print of a commonly used testing company states, "Reference ranges are representative of a healthy population *under non-challenge or non-provoked conditions.*" When Buttar described his miracle chelation treatments to a congressional subcommittee, congressmen nodded approvingly every time he showed mercury in the urine of autistic children. But the congressmen would have seen the same results had Buttar chelated them. Indeed, when researchers compared mercury excretion in children with or without autism, they found that autistic and normal children had the same amount of mercury in their bodies.

Fourth: Not only do Buttar's chelation therapies not work, but it doesn't make sense that they would. When a cell is damaged by a heavy metal such as mercury, it's permanently damaged. When doctors treat patients with chelation who really are poisoned by mercury, they do it for one reason: to bind free mercury and rid the body of it before it can do more harm. This means that when Rashid Buttar treated Desiree Jennings with intravenous chelation, claiming an almost immediate reversal of symptoms, it couldn't have been because of the chelation. Following the first reports of Desiree's remarkable recovery, Steven Novella wrote, "Brain damage does not immediately reverse itself once the cause is removed. . . . Now Jennings herself, and Dr. Buttar, report that Jennings began to improve while still sitting in the chair and receiving chelation therapy, and within thirty-six hours her symptoms were completely gone. First, let me say that I am very happy Ms. Jennings' symptoms have resolved. Hopefully now she can go on with her life. But to me, this impossibly rapid recovery is a dramatic confirmation that her symptoms were psychogenic to begin with."

I n 2009, when Rashid Buttar was asked whether he had tested his anti-autism cream to prove it worked, he responded, "No, we haven't done that. Why would I waste my time proving something that I already know works innately?"

By choosing not to test his miracle cure for autism, Rashid Buttar carries on the grand tradition of medical hucksters throughout the centuries. The claim is always the same: *It works because I know it works. It works because my patients say it works.* "This little bottle is the only thing that has been shown to conclusively get these kids better," says Buttar. Think about this for a moment. You've just invented the only medicine that you believe cures autism, a disorder that affects as many as one in eighty-eight American children. Wouldn't you be the first in line to prove that it works? To prove that it should be on the medicine shelf of every child with this disorder? When Edward Jenner thought that an injection of cowpox could prevent smallpox, he couldn't wait to test it. In 1796, Jenner proved that his vaccine worked; soon, it was used throughout the world. When Frederick Banting and Charles Best isolated insulin in 1921, they rushed to children's bedsides to prove that it worked; now insulin is standard therapy for people with diabetes, allowing sufferers to live longer. And when Howard Florey and Ernest Chain isolated, purified, and mass-produced penicillin in the early 1940s, they immediately tested it in victims of a Boston nightclub fire. So why is Rashid Buttar hesitant to test a medicine that he "knows" is the only effective treatment for autism? Probably because, once it was studied, he would have to admit that his claims are fanciful.

Desiree Jennings ultimately left Rashid Buttar, put off by the size of his bill. She shouldn't have been so surprised. One need only look at Buttar's autism treatments to see how he operates. During the first twelve months, children are required to use his chelating cream every other day, at a cost of $150 per small vial. Buttar makes sure that parents use his product only—no substitutes. "Many pharmacies are already trying to duplicate TD-DMPS by creating their own topically applied form of DMPS," he writes. "These inferiorly combined substitutes are being marketed to capitalize on our research and impersonate TD-DMPS. Proceed at your own risk."

Although sales of Buttar's anti-autism cream have been robust, they pale in comparison with Buttar's biggest seller: Trans-D Tropin, another potion he invented. As with his autism cure, claims for this transdermal drug are remarkable. And as with his autism cure, these claims have never been put to the test, which is in part why the FDA has never licensed it. "After the first few days to the first two weeks on Trans-D," writes Buttar, "most patients require less sleep and experience a better quality of sleep. . . . As time goes on, you'll experience various other changes . . . including diminished wrinkles, thicker skin, increased muscle strength and endurance, faster recovery, stronger libido, hair regrowth, increased emotional stability, higher energy levels, body contour changes and decreased chronic pain. In many instances, decades' worth of old aches, pains and injuries begin to disappear! You don't *need* to take Trans-D, but if you're interested in the possibility of increasing your life span, improving functionality and getting healthier, then you need to experience Trans-D firsthand." So, according to Buttar, Trans-D makes you look better, live longer, sleep better, and have better sex—a sales pitch that harks back to Hamlin's Wizard Oil, Squire's Grand Elixir, Kickapoo Indian

Sagwa, and other cure-alls hawked in the 1800s. Trans-D goes for about two hundred dollars a bottle. Since 1998, more than 22 million doses have been sold. Along with his anti-autism cream, Trans-D has made Rashid Buttar a very rich man.

The final irony is that while Buttar is making a fortune selling unlicensed medical products of unproven value, he rails at Big Pharma. "The motivation of most pharmaceutical companies is to fund research where they can have a monopoly," he writes, "where they can make a lot of money." Michael Specter, a staff writer for *The New Yorker*, has commented on the contradiction. "We hate Big Pharma," he says. "But we leap into the arms of Big Placebo."

Rashid Buttar asks his office staff to take an oath: "I vow to do more than my share in making the change the world is waiting for." He asks his own children to take the same oath. Buttar believes he can chelate the world into better health. It's not just a philosophy; it's a mission—a mission based on the notion that doctors are evil and that mainstream medicine can't be trusted. "Doctors often expect patients to simply believe whatever they are told," he writes. "Herein lies your first lesson. If a doctor becomes upset because you ask for more information or becomes nervous when you don't believe them simply because he or she 'said so,' you need to find yourself a new doctor. Run. Don't walk. Remember that doctors are just human beings with a license to make life-and-death mistakes as long as they are using an approved method within the 'standard of care.'"

Buttar believes that he, on the other hand, should be trusted absolutely: "I want the person who comes to me and says, 'I know what *the truth* is. I don't care about anything else.' I want you to start trusting me yesterday. That's my ideal patient. But someone who comes to me and says, 'I don't know what this is, and what are the side effects of that?' Just go your own route. I'm here for the people who already *know*." Buttar demands strict compliance with his philosophies: "There's no thinking there. I tell them I'm the general. If you want to win the race, then I have to hold the reins. And you do everything I ask you to do. If I ask you to stand on your head for four hours and chant a mantra, then you do it."

To firm up the appeal, Buttar caters to his followers' sense of conspiracy, using the catchphrase "what your doctor won't tell you." The implication is that doctors are saving the good therapies for themselves and their friends, their patients be damned. Buttar claims that as few as one in ten doctors with cancer actually get the radiation or chemotherapy recommended for them, presumably because doctors know better than to do what they recommend. And it's all part of a larger, far more heinous plot. "I was at a meeting at the Centers for Disease Control in early October," says Buttar. "And behind closed doors, I was meeting with a very senior official, a scientist, and . . . I asked what is the number-one concern for the CDC right now. And he looks at me and very pointedly he says, 'Rashid, we will deny this in public, you understand; nobody can admit to this. But the number-one concern is mercury.'"

Buttar's message is clear: *Trust me. Trust me because others mean to do you harm. Trust me because I, like you, have been treated badly. Trust me absolutely and without question. Trust me because the truth will set you free.* Buttar is in the company of charismatic figures from Jim Jones to David Koresh: building a following with unfounded, illogical notions that—in the end—benefit no one.

Still, one could argue, where's the harm? If parents want to trust Buttar's unproven tests and magical potions, if they want to buy into his logic, if they want to believe there's a conspiracy by the government to deny them important therapies like his anti-autism cream and Trans-D, and if they want to spend much of their hard-earned money doing it, that's their decision. Unfortunately, Buttar's advice is potentially quite dangerous.

Buttar's central premise is that the "medical establishment" offers unnatural and dangerous therapies. He claims that his therapies, on the other hand, are natural and harmless. Chelation therapy, however, is anything but harmless. Children who really do suffer from heavy-metal poisoning are given chelation therapy in the hospital, where their heart rhythms and blood chemistries are constantly monitored. Hospital monitoring is required because chelation medicines don't bind mercury and lead only—they also bind elements like calcium, which is necessary for electrical conductivity in the heart. In March 2006, the CDC published the stories of two children and one adult who had died from chelation. At the end of their report, CDC scientists made it clear what they thought about the unapproved use of chelation: "Certain healthcare practitioners have used chelation for autism in the belief that mercury or other heavy metals are producing the symptoms. These off-label uses of chelation therapy are *not supported by accepted scientific evidence.*"

There's another dangerous aspect to Rashid Buttar's rejection of conventional medicine. It relates to how he believed his son had become autistic. "Unbeknownst to me," writes Buttar, "my now ex-wife had gotten Abie the regularly scheduled vaccines because she had listened to the fear-evoking propaganda fed to her by the pediatricians and the doctors at the hospital when she delivered." Buttar believed that his son had been poisoned by thimerosal, a mercury-containing preservative in vaccines, saying, "Thimerosal was the greatest atrocity ever committed to mankind in the name of money." As a consequence, Buttar refused to vaccinate his third child (even though by that time thimerosal had been removed from all vaccines given to young infants) and, like Jenny McCarthy, is on a crusade to prevent others from vaccinating their children. "Nobody's giving my child any vaccine," says Buttar. "I'll take my chances with smallpox or polio or hepatitis B. Am I afraid that he will become a doctor or a prostitute by the age of ten?"

Like his argument for heavy metals as the cause of seemingly all chronic diseases, Buttar's case against vaccines is ill-founded. First, studies have clearly shown that thimerosal in vaccines not only didn't cause autism; it didn't even cause subtle signs of mercury toxicity. Next, Buttar says he will take his chances with smallpox. Fair enough. Smallpox vaccines haven't been given to children since 1972—a consequence of the disease having been wiped off the face of the earth. Polio, on the other hand, is still around, having never been eliminated from countries like Pakistan, Afghanistan, and Nigeria. People traveling from these three countries have brought the disease to twenty other countries. Given the frequency of international travel, there's every reason to believe that the virus can spread further—especially if not enough people are immunized. Finally, Buttar underestimates the impact of hepatitis B virus in children. Before the CDC recommended a routine hepatitis B vaccine for infants in 1991, about sixteen thousand children below the age of ten got infected every year. About half caught the infection from their mothers during birth, the other half after they were born, usually from casual contact with people who didn't know they were infected. As the roughly 1 million Americans infected with hepatitis B virus will attest, you don't have to be a doctor or a prostitute to get the disease.

When the United States faced a crisis in 2009, Rashid Buttar was among the first to give potentially deadly advice. "The facts are the facts," he said. "Right now more people have died from the swine flu vaccine than have died from swine flu. Probably more people will die from the swine flu vaccine than will ever die from swine flu itself. The viral strain has lost its virulence as it has come through the Yucatan Peninsula through Mexico and into the United States. So really the big hype that they made is all smoke and mirrors. It's all an illusion to scare people." Buttar was right in claiming that swine flu virus had worked its way up from Mexico, entering the United States in April 2009. But he was wrong that the virus had lost its virulence. During the few months following his pronouncements, an estimated 47 million Americans were infected with swine flu, 250,000 were hospitalized, and 12,000 died. Among the dead were an estimated 1,100 children—ten times more than die during a typical influenza season. Buttar's advice wasn't just wrong, it was spectacularly wrong. And anyone who listened to it faced an unnecessary risk.

Not all of Rashid Buttar's patients are satisfied with his care. In April 2008, the North Carolina Medical Board heard their complaints. The board's lawyer, Marcus Jimison, introduced his case. "The evidence will show that Dr. Buttar preys on people in their darkest hours, at a time when they are most desperate. He provides therapies to dying patients that have not been shown to be effective and charges them thousands of dollars a day for what he knows will not work. And he does this with a North Carolina medical license hanging on his wall."

Jimison summarized the allegations against Buttar.

- Buttar treated a patient with cervical cancer with intravenous hydrogen peroxide, an unproven and potentially dangerous therapy. For the initial visit, Buttar charged $12,000. During the next month, the patient received nineteen more injections, at a cost of $1,000 each, for a total of $31,000. When the patient died, Buttar's office sent the family a refund of $2,500.
- Buttar treated a patient with ovarian cancer with intravenous vitamins, chelation, Philbert Infra Respiratory Reflex Procedure, and Ondamed biofeedback. The bill for two months of treatments was $30,000. Prior to her death, the patient paid Buttar $10,000. When her estate failed to pay the remaining $20,000, Buttar turned the matter over to a collection agency.

- Buttar charged a patient with cancer of the adrenal gland $32,000 for ineffective therapies. The patient's wife remembered their first meeting with Buttar: "He said it didn't matter what kind of cancer anybody had," she recalled. "He could cure it. He kept reiterating he had a 100 percent success rate." After her husband died, his wife canceled a check to Buttar for $6,700. Buttar turned the matter over to a collection agency, seeking the unpaid portion of the bill, interest, and a 25 percent collection fee.
- In October 2007, Buttar told a patient with colon cancer that he "would be an idiot for doing anything the conventional doctors told him to do." Buttar advised chelation and ozone therapy, at a cost of $5,000 a week. Two months later, the patient was dead.

Toward the end of the hearing, Jimison asked Buttar whether his therapies were below the standard of care. "It's not the standard of care," he said. "It's beyond the standard of care."

Jimison closed by exhorting the medical board to do what was right—to suspend the license of a man who was doing harm, at the very least by diverting patients away from potentially helpful therapies. "Somewhere there is a loved one that's going to Dr. Buttar's office," he said. "They're dying, and would be looking for any glimmer of hope. And Dr. Buttar is more than willing to give them that glimmer of hope. This is the time to stand up for science-based, evidence-based medicine. . . . It will be the board's finest hour."

It wasn't to be. The board chose only to order Buttar to provide a consent form advising his patients that his treatments hadn't been proven effective and hadn't been licensed by the FDA. He could continue to treat patients with cancer and autism. Continue to treat them as if they had been poisoned by heavy metals, even though evidence refuted his claim. Continue to sell magical anti-autism creams and anti-aging medicines without any proof that they worked. Buttar still describes treatments on his website as "highly effective" and boasts of giving 500,000 chelation treatments without any side effects (which is highly unlikely). It would be surprising if the revised consent deterred desperate patients and parents from streaming to his door.

Part VII

WHY SOME ALTERNATIVE THERAPIES REALLY DO WORK

The Remarkably Powerful, Highly Underrated Placebo Response

We are what we pretend to be ...
—Kurt Vonnegut, *Mother Night*

So if echinacea and vitamin C don't treat colds, and chondroitin sulfate and glucosamine don't treat arthritis, and St. John's wort doesn't treat depression, and ginkgo doesn't improve memory, and saw palmetto doesn't shrink prostates, then why do so many people believe they do? And if the human nervous system isn't related to rivers in China, and all diseases aren't based on misaligned spines, and highly diluted medicines don't contain any active ingredients, then why do acupuncturists, chiropractors, and homeopaths have such a devoted following? The answers can be found during a testy exchange between Mehmet Oz and Steven Novella.

In April 2011, Oz discussed acupuncture on *The Dr. Oz*

Show. Researchers have spent a lot of time and money studying acupuncture in people who claim it works. First, they compared outcomes when needles were inserted into correct or incorrect acupuncture points. No difference. Then, they used both standard and retractable needles; patients felt the sting of the needle but didn't know whether it had entered the skin. Again, no difference. Despite these studies, Mehmet Oz continues to promote acupuncture. To his credit, however, on the day he discussed acupuncture, Oz invited Dr. Steven Novella onto his show. He couldn't have picked a more skeptical guest.

In addition to being a Yale neurologist, Steven Novella is founder and president of the New England Skeptical Society, director of the Science-Based Medicine project at the James Randi Educational Foundation, and host of the popular science podcast *The Skeptics' Guide to the Universe*. Novella explained that acupuncture still works if needles are inserted in the wrong place or if they aren't inserted at all. Oz was furious. "There are billions of people around the world who use acupuncture as the foundation of their health," he argued. "I just think it's very dismissive of you to say that because we couldn't take this idea that exists with a different mind-set and squeeze it into the way that we think about it in the West, that it can't possibly be effective."

Novella knew that acupuncture was by definition a sham, a trick, a deception; yet he never once said, "Acupuncture doesn't work." Rather, he questioned *why* it worked. "It's the ritual surrounding a positive therapeutic interaction: a comforting, caring [clinician]," he said. "You're relaxing for half an hour or an hour. That's where the effect is. There's no effect to actually sticking a needle through the skin." In other words, the placebo effect.

Although some dismiss the placebo effect as trivial, it's not. One of the first demonstrations of how powerful it can be took place on the battlefields of World War II, when a nurse ran out of morphine. Unable to tell a wounded soldier that she had nothing for his pain, she lied, saying that the salt water she used was actually morphine. To her surprise, his pain disappeared. The critical question, then, isn't "Does the placebo effect work?" It's "How does it work?" Do placebo therapies such as acupuncture really cause less pain, or do people simply tolerate the same pain? Is the placebo effect physiological or psychological?

At first, doctors were unfairly dismissive of the placebo effect, arguing that it had everything to do with perception and nothing to do with reality. For example, expensive therapies are often perceived as more valuable even if they're worthless. One therapy that falls into this category is unicorn horns. Touted for the treatment of epilepsy, impotence, worms, plague, smallpox, and rabies, and purported to prolong youth, assist memory, and fortify the spirits, powdered unicorn horns have been used for more than eight centuries. Unicorns are one-horned beasts that don't exist—a fact that hasn't hurt sales. Made from ground-up whales' tusks, "unicorn horns" sell for their weight in gold: a nine-pound horn sells for $55,000. People who can't afford unicorn horns can buy unicorn drinks.

A more recent example is Vitamin O. In 1998, an advertisement in *USA Today* announced a miraculous new product: Vitamin O. Under a photograph of tensely smiling, attractive people, the ad stated, "It's so safe you can drop it in your eyes, so natural it contains the most abundant element on earth, so effective you can spend hours reading the unsolicited testimonials of those who've used it with dramatic results." One believer said, "After taking Vitamin O for several months, I find I have more energy and stamina and have become immune to colds and flu." What was Vitamin O? The ad didn't lie. Vitamin O was "stabilized oxygen molecules in a solution of distilled water and sodium chloride." In other words: salt water. But promoters of Vitamin O were quick to point out that it wasn't *just* salt water. It was salt water enhanced with oxygen, giving buyers the vital energy they needed. Unfortunately, Vitamin O users lacked the one thing necessary to extract oxygen from water: gills. After the ad appeared, the manufacturer sold sixty thousand vials in a month—at a cost of $20 a vial.

Psychologists have also argued that the placebo effect is simply an exercise in conflict resolution. People who use acupuncture are confronted with two conflicting facts: (1) acupuncture is unconventional; (2) acupuncture is expensive, costing between $65 and $120 a session, requiring frequent sessions, and often necessitating out-of-pocket payments. The conflict is best resolved by believing that acupuncture works. In his book *When Prophecy Fails*, Leon Festinger called it the "theory of cognitive dissonance," the best example of which is Aesop's fable "The Fox and the Grapes." A fox comes upon a bunch of grapes hanging from a tree and confronts two opposing thoughts: (1) he loves grapes; (2) he can't reach the grapes. The fox resolves the conflict by convincing himself that the grapes are sour.

Another explanation for the placebo effect is something

called "regression to the mean," a phenomenon best exemplified by the *Sports Illustrated* cover jinx. For many years, sports fans have argued that an appearance on the cover of *Sports Illustrated* is the kiss of death for a professional athlete. The story is always the same: an athlete has a great season, gets on the cover of the magazine, then has a mediocre or poor season. But this isn't really a jinx. When an athlete has a great season, it shouldn't be surprising that the next season wouldn't be as great. That's because athletes have up and down years. The same can be said for pain, which can wax and wane. People usually go to an acupuncturist when they're experiencing the most pain. So it's possible that they feel better after their worst suffering because pain, like an athlete's career, can fluctuate.

During his face-off with Mehmet Oz, Dr. Novella alluded to yet another explanation for the placebo effect: the therapist's presentation. Many studies have shown that a therapist's clothes, demeanor, attitude, and phrasing make a difference. Therapists who say, "You will be better soon" or "These pills will help" find that patients do better than those who say, "I don't know what you have" or "I don't know if pills will help." And the more time a therapist spends with patients, the better they do. Ted Kaptchuk, an alternative healer from Cambridge, Massachusetts, recognizes the healing power of personality. "I am a damn good healer," he says. "That is the difficult truth. If you needed help and you came to me, you would get better. Thousands of people have. Because, in the end, it isn't really about the needles. It's about the man." "The doctor who fails to have a placebo effect on his patients," wrote J. N. Blau, "should become a pathologist."

Probably the best example of a therapist's influence can be found in *The Wizard of Oz*. Because the Wizard can't give the

Scarecrow a brain, he does the next best thing: he makes him *feel* smarter. "Back where I come from, we have universities," says the Wizard. "Seats of great learning where men go on to become great thinkers. And when they come out, they think deep thoughts—and with no more brains than you have. But! They have one thing you haven't got! A diploma!" After receiving the diploma, the Scarecrow is smarter, reciting the Pythagorean theorem. "Oh joy, rapture," he says. "I've got a brain."

In a sense, everyone uses the placebo effect, no one more than parents. "I have a four-year-old son," writes John Diamond in *Snake Oil and Other Preoccupations*, "who, as is the way with four-year-old sons, climbs on things and then falls off them, losing chunks of skin and blood on the way down to the ground. And whenever this happens—about four times a week on average—and he runs to me crying and pointing to the latest graze, I apply strictly alternative remedies. I don't give him strong drugs to kill the pain or stanch the flow of blood but I clean the tiny wound, ask him what happened, sympathize with him about the fickleness of gravity and the harshness of brick, sit him on my lap with a glass of water, kiss him and rub the hurt better. And it works, every time." Diamond tried to label his not-so-unique brand of healing. "So what do you think I should call the technique?" he writes. "Fatherotherapy? Dynamic parentesiology? Because dealing with minor problems by calm talking, sitting down quietly and rubbing it better seems to be just what much of alternative therapy is about."

B y the 1970s it had become clear that psychologists hadn't appreciated the whole story. And that what alternative healers had been calling "the mind-body connection" had a physiological basis. That's when the *placebo effect* became the *placebo response*. It wasn't that people merely *believed* they had less pain; they *did* have less pain. Dismissive notions that pain relief was "all in their heads" were replaced by a better understanding of how pain works. And why the placebo response cannot and should not be ignored.

In the 1990 movie *Postcards from the Edge*, two women (Evelyn and Suzanne) realize they've been sleeping with the same man. Suzanne is mortified; Evelyn isn't.

SUZANNE: When did you see Jack last?

EVELYN: Umm, Saturday. Saturday night.

SUZANNE: I was with him Saturday afternoon. That's two girls in one day.

EVELYN: And that's just the ones we know about. Think what you could find out if you had one of those satellite things.

SUZANNE: How can you laugh? It's completely disgusting. Especially in this day and age.

EVELYN: You look like someone who can take care of herself. Buy some condoms. Don't feel bad. He probably really likes you. If you can just . . . enjoy yourself with him like he's enjoying himself with you. That's what I do. I'm in it for the endolphin rush.

SUZANNE: Endorphin.

EVELYN: Whatever.

Evelyn had offered a clue to the physiological basis of the placebo response.

The single most powerful pain-relieving drug is morphine. Isolated from the poppy plant (*Papaver somniferum*), it's the analgesic to which all others are compared. In the early 1970s, Rabi Simantov and Solomon Snyder discovered the receptor in the brain that binds morphine. That wasn't surprising. What was surprising was that they also found chemicals that acted just like morphine and bound to the morphine receptor. These chemicals weren't derived from plants or synthesized by pharmaceutical companies; they were made by the human pituitary gland and hypothalamus. Simatov and Snyder called them endorphins, a contraction of *endogenous* (produced in the body) and *morphine*. Later it was shown that endorphins are released in response to pain, spicy foods, exercise, excitement, and orgasm (Evelyn's "endolphin rush").

With endorphins in hand, scientists could better determine how remedies like acupuncture worked. In 1978, Jon Levine, Newton Gordon, and Howard Fields divided people who had dental surgery into two groups. Both groups received diazepam (Valium), nitrous oxide ("laughing gas"), and a local nerve block (mepivacaine). After these analgesia wore off, one group also received a placebo pill to "relieve the pain" and the other group received nothing. Many in the placebo group experienced pain relief. More important, the placebo response was eliminated by administration of naloxone, a chemical that blocks endorphins. In a paper titled "The Mechanism of Placebo Analgesia," the authors concluded, "These data are consistent with the hypothesis that endorphin release mediates placebo analgesia for dental postoperative pain." Other groups have reproduced these findings.

The results were in. Placebo pain relief could be physiological, real. When people said they felt less pain after acupuncture, it wasn't "all in their heads"; it was in their bodies, caused by the body's own drug. (Researchers have also shown that pain relief from acupuncture can be blocked by naloxone.) Knowing this, alternative healers argued that if acupuncture can spare people from prolonged use of analgesics—some of which have significant side effects—what's the harm? Wouldn't it be better to induce endorphins naturally than to rely on drugs?

Those who dismiss acupuncture make three counterarguments.

First: acupuncture is a deception. If acupuncturists were honest about studies comparing real to sham acupuncture, they'd say to their patients, "Before we begin, let's set aside all this two-thousand-year-old-ancient-wisdom business. The truth is the Chinese didn't believe in dissection, knew nothing about the anatomy of the nervous system, wrongly assumed it was based on rivers in China and the lunar month, and inserted needles under the skin randomly. Forget *chi*; forget *yin* and *yang*; forget meridians. Acupuncture will work just as well if I use retractable needles. The reason it works is that you *think* it works. And thinking alone might be enough to release endorphins." Acupuncturists don't say this for the obvious reason that it would probably eliminate the placebo response. Maybe it would eliminate it because the mental image of *yin*, *yang*, and *chi* are important to the therapeutic process. Or maybe patients respond better to the assurance of ancient wisdom than to the caveats of modern science. Whatever the reason, deception is likely an essential part of the therapy.

Art Caplan, professor of bioethics at New York University's Langone Medical Center, has addressed the ethics of placebo medicine. "It's ethical to deceive the patient at low risk, at low cost, and at low burden," he says. "Medicine can learn from chiropractors, can learn from acupuncturists. But they have a duty to report what they're doing in the medical literature. They should report that the placebo effect is powerful, that certain things can induce it, and that medicine ought to study how it can best be elicited."

In fairness, all practitioners—mainstream or otherwise—employ some form of deception. They know that a positive attitude, reassuring demeanor, and air of competence are important. "We use the placebo effect all the time," says Caplan. "I've got a bow tie. I wear a white coat. You come to a big building that looks pretty impressive. I expect someday to see billboards go up in cities that say we have a really big machine and it makes a lot of noise and we don't know how it works, but you can only get it from us, so come on down."

Indeed, it would be more honest if mainstream doctors walked into a patient's room and said, "Look, we will definitely know more about how to treat you a hundred years from now. Frankly, I suspect doctors in the future will look back on some of the things we're doing today and laugh. Although our understanding of many diseases is excellent, for some we're just treading water, and for others we're completely lost." No clinician (in his right mind) says this. From the days of shamans and witch doctors to the modern-day physician, everybody has their props, their deceptions.

The second argument against acupuncture is that it's expensive. But, at least according to the theory of cognitive dissonance, the more expensive, the better. This concept was first tested at a racetrack in the 1960s. Researchers asked bettors to rate their horse as they walked toward and away from the betting window. Bettors faced two conflicting facts: (1) any horse could win the race; (2) I bet a lot of money on only one horse. To resolve the conflict, bettors rated the horse they'd picked much higher *after* placing their bets. The study was titled "Post-Decision Dissonance at Post Time." In another study, researchers from MIT tested the capacity of two sugar pills to relieve pain. One group was told that the pill cost 10 cents, the other that it cost $2.50. Participants experienced less pain with the more expensive pill. "Look, at the end of the day, if you say to me the only way this works is if we charge fifty bucks and pretend we are going through this ceremony, and that's the only thing that gives pain relief—okay," says Caplan. "But I don't think we're there yet. I don't think we've fully explored how to elicit the placebo response in other ways."

The final argument against acupuncture is the hardest to refute. Acupuncture needles are not without risk, having punctured hearts, lungs, and livers and transmitted viruses like HIV, hepatitis B, and hepatitis C. Perhaps the most famous case involved the former president of South Korea, Roh Tae-woo, who had an acupuncture needle removed from his lung in May 2011. "I can't figure out how the needle got into there," said Dr. Sung Myung-whun, the operating surgeon. "It's a mystery for me, too." At least eighty-six people have died from acupuncture.

If sham acupuncture works as well as real acupuncture, and if putting needles under the skin can puncture lungs and cause infections, why not use retractable needles? Acupuncturists might argue that the use of retractable needles would be a deception; but they're already knee-deep in the deception that acupuncture points have anything to do with the nervous system. So what's one more? The goal should be to induce endorphins in the safest way possible.

The discovery of endorphins changed everything. Now there was a clear, rational, physiological mechanism by which therapies that weren't anchored in the anatomy of the nervous system could work. But alternative practitioners don't limit themselves to chronic pain. They offer relief from a variety of immunologic, neurologic, and metabolic diseases, again with therapies that are often unrelated to the physiological basis of those diseases. Like the identification of endorphins, another discovery in the 1970s shed light on why some of these therapies might work. The discovery was so surprising that, until it was reproduced in two other laboratories, no one believed it.

In 1975, Robert Ader and Nicholas Cohen, from the University of Rochester School of Medicine, published a paper titled "Behaviorally Conditioned Immunosuppression." The experimental design was simple. Ader and Cohen injected one group of rats with sheep red blood cells; as expected, the rats made antibodies against the cells. A second group was given saccharine-flavored water at the same time they were injected with the cells; like the first group, they developed an immune response. A third group of rats was inoculated several times with the cells suspended in saccharine-flavored water containing cyclophosphamide—an immunosuppressive drug—which, not surprisingly, inhibited the immune response. Then Ader and Cohen gave this third group the red blood cells in saccharine-flavored water *without* cyclophosphamide. And they found something remarkable: *saccharine-flavored water alone suppressed the immune response*. By pairing an immunosuppressive drug with a distinct taste, the third group of rats had *learned* to suppress their own immune systems. Amazing.

In a way, the Ader-Cohen experiment shouldn't have been that surprising. In 1896, J. N. MacKenzie studied several people who suffered itching, sneezing, and watery eyes when exposed to pollen on flowers—an allergic response mediated by histamine. In a paper published in the *American Journal of the Medical Sciences*, MacKenzie reported that artificial flowers elicited the same symptoms, even though they were pollen-free. People had learned to make an allergic response—learned to release their own histamine.

In 1957, John Imboden, Arthur Canter and Leighton Cluff, scientists at Johns Hopkins School of Medicine, performed another landmark experiment. They administered a series of psychological tests to military personnel working at Fort Detrick, in Frederick, Maryland. A few months after the tests were completed, an influenza pandemic swept across the camp. Imboden and his colleagues found that recruits who were depressed had symptoms of influenza that lasted longer and were more severe than those who weren't. Mood determined illness. The results of this study lent credence to the adage that people get sick when they want to get sick. "The mind," wrote Milton in *Paradise Lost*, "can make a heaven of hell and a hell of heaven."

The next question was, could these findings be put to practical use? Could researchers teach people to suppress or enhance their own immune response? Robert Ader was one of the first to step forward. Working with a teenager with the autoimmune disease lupus, Ader paired cyclophosphamide with a distinct taste (cod liver oil) and smell (rose perfume). Like the rats, the boy learned to suppress his immune response, requiring less frequent dosing of the drug needed to control his disease. Others replicated Ader's findings. Marzio Sabbioni found that healthy men could learn to release their own cortisol, a natural steroid produced by the adrenal gland. And it worked both ways: not only could people learn to suppress their immune responses; they could also learn to enhance them.

If people can learn to stimulate or suppress their own immune responses, it's not a leap to believe that placebos can impact a variety of diseases. Even though most alternative medicines don't work better than placebos, some placebos work. So why not use them?

For example, Wolf Storl's claim that the weed teasel cures Chronic Lyme disease might be of some value. Even though Chronic Lyme disease doesn't exist, chronic pain and fatigue do. For some, it's possible that the mental image of bacteria being killed by teasel causes them to experience less pain and fatigue. The same can be said for the image of balancing *yin* and *yang* to release *chi*. If these mental constructs work to relieve pain, where's the harm? Teasel's cheap and is better than taking long-term pain medications (or, worse, long-term antibiotics). Similarly, Bryan Rosner's promotion of Rife machines is harmless. Although claims that it kills Lyme bacteria are fanciful, at least it doesn't kill anything else. (The electric current produced by the machine probably doesn't penetrate the skin.) Assuming that the problems aren't amenable to conventional therapies, one could make the same argument for treating chronic symptoms with magnets, crystals, saunas, aromatherapy, emu oil, or prayer.

Furthermore, many alternative healers recommend chondroitin sulfate and glucosamine for chronic joint pain. Although these remedies don't work better than placebo pills, they're harmless. And if that means avoiding pain medicines that occasionally cause serious side effects, why not give them a try? Even though chondroitin sulfate and glucosamine don't work better than placebos, that doesn't mean that they don't work as placebos.

Another example of the potential value of placebos is homeopathic remedies.

In March 2011, Mehmet Oz invited a homeopath named Russ Greenfield onto his show. Oz began by saying that his wife gives homeopathic medicines to their children all the time. "It's only when they don't work that she calls me," he said. Homeopathy is based on the notion that if you take a particular medicine and dilute it to the point that not a single molecule remains, the water will "remember" that the drug was there. (Given the finite amount of water on earth, it's comforting to know that water doesn't really remember where it's been.) Greenfield explained that the "spirit of the medicine was there." A win-win situation. People can enjoy the benefit of a drug without having to suffer any of its side effects (since it's not there anymore).

Oz described homeopathy as medicine present in "very small diluted amounts"—so dilute, you can "hardly find a molecule."

He showed his viewers two large jugs of fluid. One, a frightening red color, was labeled "Conventional Medicine." The other, containing only water, was labeled "Homeopathic Medicine." Oz filled a dropper with red fluid from the "Conventional" jug and added it to the "Homeopathic" jug, which turned faintly pink. This, he declared, was the basis of homeopathy. Any reasonable viewer would assume that homeopathic medicines are weak preparations of conventional medicines. But they're not. It would have been more accurate if Oz had taken a dropper full of air, added it to the "Homeopathic" jug, and said that homeopathic medicines don't contain any medicine at all. Not surprisingly, homeopathic remedies haven't been found to be more effective than sugar pills or water. (After all, they *are* sugar pills or water.) Five excellent studies have shown that homeopathic remedies didn't treat cancer, attention deficit/hyperactivity disorder, asthma, dementia, or flu-like symptoms. But again, just because a drug doesn't work better than a placebo doesn't mean it doesn't work. It only means that it isn't better than the placebo response.

For example, one popular homeopathic remedy is oscillococcinum, promoted for the treatment of flu. Homeopaths make oscillococcinum by homogenizing the heart and liver of a Burberry duck, diluting it in water one-hundredfold, and repeating the hundredfold dilution two hundred times. A solution this dilute doesn't contain a single molecule of the Burberry duck. In fact, the preparation is so dilute that not a single molecule of the duck would be found if the final volume were that of the universe. The duck is gone. From a scientist's standpoint, oscillococcinum is one gram of sugar.

Although oscillococcinum is inert, it's a lot better than other remedies out there, many of which are quite harmful. In 2007, the FDA issued a warning against the use of cough-and-cold preparations in young children—a warning long overdue. Between 2004 and 2005, the CDC found that more than fifteen hundred young children had suffered hallucinations, seizures, and heart problems caused by cough medicines containing stimulants like pseudoephedrine. Three children had died as a result. Worse: cough-and-cold preparations do little to relieve symptoms.

Oscillococcinum, on the other hand, doesn't cause any of these problems. And although it works no better than a gram of sugar, many people—possibly benefiting from the placebo response—believe that it does. Most important, it avoids the harm caused by medicines that are all too available over the counter.

Although these therapies are often referred to as "New Age" medicine, there's nothing new about them. Healers have been selling the placebo response for five thousand years. The Ebers Papyrus, written in 1500 B.C., contains more than seven hundred drugs of mineral, vegetable, and animal origin. Treatments included dirt and flyspecks scraped off walls; blood from lizards and cats; fat from geese, oxen, snakes, mice, hippopotamuses, and eunuchs; dung from pigs, dogs, sheep, gazelles, pelicans, and flies (this couldn't have been easy); and ram's hair, tortoise shells, swine's teeth, donkey's hooves, and grated human skulls. Again, just because pelican poop doesn't contain real medicine doesn't mean it didn't work to treat pain, fatigue, and heartburn 3,500 years ago. The placebo response is a powerful thing. And back then it was all they had. Now, because we have medicines that are much better than placebos, we often ignore the placebo response. To their credit, alternative practitioners haven't. It would be of value, however, if they could learn to do it without the pills, needles, electrical devices, and appeals to magical thinking.

One of the greatest proponents of alternative medicine was Norman Cousins, longtime editor of the *Saturday Review*. Cousins suffered from joint and muscle pain, which he believed was cured by his indomitable will to live and laughing at Marx Brothers movies. ("There ain't much fun in medicine," wrote Josh Billings, an American humorist, "but there's a heck of a lot of medicine in fun.") Like Art Caplan, Cousins reasoned that the placebo response could be elicited *without* placebos. He wrote about his experiences in *Anatomy of an Illness*, a national best seller in the late 1970s: "The placebo is only a tangible object made essential in an age that feels uncomfortable with intangibles, an age that prefers to think that every inner effect must have an outer cause. Since it has size and shape and can be hand-held, the placebo satisfies the contemporary craving for visible mechanisms and visible answers. But the placebo dissolves on scrutiny, telling us that it cannot relieve us of the need to think deeply about ourselves. The placebo, then, is an emissary between the will to live and the body. But the emissary is expendable."

12

When Alternative Medicine Becomes Quackery

. . . So we must be careful about what we pretend to be.
—Kurt Vonnegut, *Mother Night*

Although alternative therapies can be valuable, a sharp line divides those who practice placebo medicine from those who practice quackery. The line is crossed in four ways.

First: by recommending against conventional therapies that are helpful.

For example, Mehmet Oz's publicizing Dr. Issam Nemeh's claim that faith healing can cure cancer was wildly irresponsible. Many children have died because their parents believed that prayer (not antibiotics) cured bacterial meningitis; or that prayer (not insulin) cured diabetic coma; or that prayer (not radiation or chemotherapy) cured lung masses.

Andrew Weil has also occasionally steered people away from effective therapies. In 1995, Weil wrote, "Western medicine's powerlessness against viral infections is clearly visible in its ineffectiveness against AIDS. Chinese herbal therapy for people infected with HIV looks much more promising." Weil's recommendation was unfortunate. Eight years earlier, antiviral drugs had been shown to decrease HIV replication, lower the amount of HIV in the bloodstream, and prolong the lives of AIDS sufferers. Chinese herbs, on the other hand, haven't been proven to do any of those things. Patients who choose Chinese herbs instead of antiviral medicines are making a choice to shorten their lives.

Asthmatics have also suffered. In July 2011, researchers at Harvard Medical School gave a group of thirty-nine asthmatics albuterol (a bronchodilator), a placebo, sham acupuncture, or nothing. Albuterol dilated breathing passages much better than placebos. Alternative healers who recommend placebos instead of bronchodilators for severe asthma are putting their patients at unnecessary risk. In 2006, a six-year-old boy with severe asthma was treated with a homeopathic remedy instead of the bronchodilator that would have saved his life. Homeopaths have recommended their products for other treatable diseases, such as cancer, malaria, cholera, and AIDS.

Chiropractors also put patients at unnecessary risk when they attempt to treat medical illnesses. For example, a chiropractor named Marvin Phillips convinced the parents of Linda Epping, a little girl with a tumor of her eye, that surgery was unnecessary. All she needed was to "chemically balance" her body with vitamins, dietary supplements, and laxatives. Within three weeks, the tumor was the size of a tennis ball: too late for surgery. In three months, Linda Epping was dead.

But when it comes to steering people away from modern medicine, perhaps no one has been more irresponsible than another of Mehmet Oz's "Superstars": Joe Mercola, an osteopath from suburban Chicago. Mercola is against pasteurization (heating at 145 degrees Fahrenheit for thirty minutes), claiming that it makes milk less nutritious. In fact, pasteurization doesn't destroy critical nutrients in milk. It destroys bacteria like salmonella, *E. coli*, campylobacter, and listeria, which can cause severe and occasionally fatal infections. About two hundred people get sick every year in the United States from drinking raw milk or eating unpasteurized cheese.

Mercola also believes that vaccines are dangerous and should be avoided or delayed. In April 2011, he helped pay for an advertisement on the CBS JumboTron in Times Square promoting his website as well as that of a prominent anti-vaccine group. And he questions whether HIV causes AIDS. "Exposure to steroids and the chemicals in our environment, and the drugs used to treat AIDS, stress, and poor nutrition are probably the real causes," he argues.

Furthermore, Mercola suggests that cancer can be treated with baking soda and coffee enemas, and he warns against mammography, offering his own special device to detect cancer, inflammation, neurological problems, and vascular dysfunction (for which he has received a warning letter from the FDA). On *The Dr. Oz Show*, Mercola said, "our model doesn't kill anyone," which would be true, except that people can die from vaccine-preventable diseases or infections caused by unpasteurized products or denying that HIV is the cause of AIDS or treating cancer with baking soda and coffee enemas instead of medicines that work.

The second way that alternative healers cross the line into quackery is by promoting potentially harmful therapies without adequate warning.

Andrew Weil writes that kava (*Piper methysticum*) "relaxes muscles, promotes calm, and is non-addictive." Seven years later, when kava was shown to cause severe liver damage, the FDA issued a warning against its use. Indeed, few consumers are aware of possible side effects from many supplements easily purchased over the counter.

Chiropractic manipulations also aren't risk-free. In the 1980s, a chiropractor named J. Richard Stober popularized a technique called Bilateral Nasal Specific (BNS) therapy. To properly align bones of the skull, Stober stuck balloons into the nose. In 1983, a baby died of asphyxiation from BNS. In 1988, school officials in Del Norte County, California, subjected children with epilepsy, Down syndrome, cerebral palsy, dyslexia, and other learning disorders to a technique developed by a New York City chiropractor named Carl Ferreri that involved squeezing the skull with both hands like a vise. Chiropractors also pressed their thumbs hard against the roof of the mouth and the eyes. One chiropractor, who did most of the treatments, said they "removed static from the nervous system." Children were forcibly restrained as they cried, screamed, and struggled to free themselves. One experienced a seizure while he was having his eye "adjusted." In 2010, Edzard Ernst, a professor of complementary medicine at the University of Exeter, in England, reviewed the medical literature and found twenty-six deaths from chiropractic manipulations, almost all caused by ripping the vertebral artery in the neck.

The third way in which alternative healers cross the line is by draining patients' bank accounts, probably best evidenced by Stanislaw Burzynski's willingness to charge the parents of Billie Bainbridge £200,000 for the false hope of antineoplastons.

But when it comes to cashing in on alternative medicine, few have benefited more than Mehmet Oz's "Superstars," all of whom have done quite well for themselves. Andrew Weil has his own brand of supplements under the name Dr. Weil's Select. Visitors to Weil's website can buy his Memory Support formula, containing ginkgo, for $56.10, his Joint Support formula, containing glucosamine, for $72, and his Energy Support formula, containing large quantities of vitamins, for $72.60. In 2003, Weil signed a deal with Drugstore.com that paid his company $3.9 million.

Deepak Chopra uses a different business model. He promotes ayurvedic supplements, oils, massages, and herbs under the brand name Chopra Center. He also sells books, videos, clothes, aromatherapies, jewelry, gifts, and music. Customers can attend courses run by Dr. Chopra titled "Journey into Healing," "Free to Love, Free to Heal," and "Summoning the Sacred," for between $1,000 and $5,000 each. If customers want to participate in a sacrificial ceremony to please the Vedic gods, it costs between $3,000 and $12,000. One year of anti-aging medicines can cost as much as $10,000. If CEOs want Deepak Chopra to speak at a corporate event, it costs $25,000.

(The Atlantic Richfield Company has employed Chopra for more than a decade.) Chopra's center grosses about $20 million a year.

Because Chopra charges a lot of money for his medicines and seminars, he has to appeal to a well-heeled clientele. At this he is a master. Chopra has helped wealthy businessmen, celebrities, and politicians like Michael Jackson, Elizabeth Taylor, Winona Ryder, Debra Winger, Madonna, Mikhail Gorbachev, Michael Milken, and Hillary Clinton find their inner space without feeling bad about being rich. Chopra writes that he "will give you the ability to create unlimited wealth with effortless ease" and that "money is really a symbol of the life energy." He has written a book titled *Golf for Enlightenment*. Chopra's unusual marriage of ancient Indian healing with unbridled capitalism is easily parodied. In 1998, in response to Chopra's *The Seven Spiritual Laws of Success*, Christopher Buckley, John Tierney, and Brother Ty wrote *God Is My Broker: A Monk Tycoon Reveals the 7½ Laws of Spiritual and Financial Growth*, which includes advice such as "God loves the poor but that doesn't mean he wants you to fly coach," "As long as God knows the truth, it doesn't matter what you tell your customers," and "Money is God's way of saying thanks." The book also includes "A Prayer for Untroubled Wealth."

Joe Mercola is also a phenomenal salesman. *BusinessWeek* described Mercola's aggressive marketing practices as giving "lie to the notion that holistic practitioners tend to be so absorbed in treating patients that they aren't effective businesspeople. While Mercola [claims to be free] from 'all the greed-motivated hype out there in healthcare land,' he is a master promoter, using every trick of traditional and Internet direct marketing to grow his business . . . an unfortunate tradition made famous by old-time snake oil salesmen of the 1800s."

Mercola sells:

- A one-ounce bottle of Organic Sea Buckthorn Anti-Aging Serum for $22.
- A pack of sixteen "worry free" organic cotton tampons for $7.99, arguing that your tampon "may be a ticking time bomb."
- The Mercola Vitality Home Tanning Bed for $2,997 ("Incredible Deal!")—remarkable, given that UV radiation is a known cause of skin cancer.
- An eight-ounce bottle of Natural Sunscreen with Green Tea for $15.97.
- And krill oil, probiotics, coenzyme Q10, multivitamins, astaxanthin, "Joint Formula," protein powder, "Purple Defense," vitamin K, chlorella, vitamin B12 spray, vitamin D spray, melatonin sleep spray, "eye support," "immune support," "prostate and bladder support," organic skin care, bath care, organic body butter, organic cocoa cassava bars, whey protein bars, tulsi teas, ceramic cookware, shampoo and conditioner, "cardiovascular support," acai berry, BioTHIN, CLA supplement, digestive enzymes, wool bedding, "dental care," Greener Cleaner, water filters, Soft Spray Bidet, pet products, fitness products, summer survival kit, tanning systems, Fiber Harmony with psyllium, herbal supplements, Chyawanprash herbal jam, "New Start," pure gold raw honey, royal matcha green tea, coconut oil, vitamin E, Himalayan salt lamps, Himalayan cooking and bath salt,

fish oil, D-Mannose, Healthy Chef sixteen-piece cutlery sets and Turbo Ovens, lightbulbs, air purifiers, juicers (even though he recommends against fruit juices), fruit and vegetable wash, kefir and vegetable starters, ferrite beads, coconut flour, sleep masks, a sun alarm clock, and massage tools, most under his own brand name.

Mercola's products gross about $7 million a year. Ironically, when asked about criticism from mainstream physicians and warning letters from the FDA, he responded, "It's very simple. There's this huge collusion between government and industry." This from a man who has *become* an industry. When Mehmet Oz questioned Mercola about his sales tactics on *The Dr. Oz Show*, Mercola said, "We only sell natural products," which presumably justifies hefty profits based on unsupported, potentially dangerous claims.

Sadly, mainstream physicians are also cashing in on the alternative medicine craze. Many academic institutions now have divisions of alternative, integrative, and holistic medicine, and the medical marketplace is starting to resemble an Arabian bazaar. "The more medicine nuzzles up to non-evidence-based alternative products," says Art Caplan, "the more you damage medicine in the long run. Many doctors believe a false premise that the patient is king. It's what I sometimes refer to as the restaurant model of medicine. Well, medicine isn't a restaurant, and the patient is not a patron, and the doctor is not a waiter. I understand that medicine is a business and that alternative medicine is a business; but part of the difference in medicine is that we're supposed to have professional norms, professional values, and professional commitments. If you keep telling people that it's just a marketplace and that they're just clients and that autonomy of the patient is what must be served to make them happy customers, then you have a collapse of professionalism in the face of consumer demand. We have this image that patients aren't vulnerable. But they are. It's hard to be a good patient unless you have an advocate who's guided by professionalism to help you. Otherwise, there are plenty of healers who will rip you off."

The fourth way alternative healers cross the line is by promoting magical thinking, which, sadly, is everywhere you look. For example, take America's favorite pastime.

In addition to their uniforms, gloves, high socks, and cleats, many baseball players wear something else: titanium necklaces. Made by Phiten, a Kyoto-based company, titanium necklaces are purported to cause "longer lasting energy, less fatigue, shortened recovery time, and more relaxed muscles." How do they work? "Everybody has electricity running through their bodies," explains Scott McDonald, a Seattle-based sales representative. "This product stabilizes that flow of electricity if you're stressed or tired. Pitchers are seeing that they aren't as sore. Injured players are seeing that they recover faster from their workouts. People are always skeptical, but when they try it, they become believers." "I think I have a little more energy with it," said Endy Chávez, an outfielder on the Seattle Mariners.

Electricity is the flow of electrons. Flow in one direction is called direct current; in two directions, alternating current. For electrons to flow, they need to be displaced from atoms. Some metals, like gold, silver, and copper, give up their electrons easily (and are therefore excellent conductors of electricity). Others, like titanium, hold on to their electrons more tightly. Independent of the metal, one fact about electricity is clear: electrons *cannot be displaced without force*, usually in the form of chemical energy (batteries) or mechanical energy (generators). Although titanium, like most metals, can conduct electricity to some extent, it cannot generate electricity without force. Absent this force, titanium necklaces are inert, no different than wearing a necklace made of wood or garlic. Perhaps most amazing is that these same athletes routinely climb into MRI (magnetic resonance imaging) machines. The magnet in one of these machines generates *sixty thousand times* the energy found in the earth's magnetic field. If titanium necklaces evoked the kind of energy athletes believe, an MRI would make their heads explode.

Titanium necklaces can be purchased on the Internet for about $40.

Unfortunately, magical thinking isn't harmless. There's a price to pay. "I don't want people running around saying that because I waved a black cat in front of them that black cats have curative power," says Caplan. Robert Slack, writing for the Center for Inquiry, agrees: "The gaps in medical knowledge we all dread are not likely to be filled by energy fields, meridians, and astrology but by the purposeful pursuit of knowledge under a single set of standards we call science. The way forward is through a careful and purposeful pursuit of scientific truth, even if it means leaving some of our most romantic fallacies behind." Isn't it enough to see that a garden is beautiful," wrote Douglas Adams, author of *The Hitchhiker's Guide to the Galaxy*, "without having to believe that there are fairies at the bottom of it, too?"

Encouragement of scientific illiteracy—or, beyond that, scientific denialism—can have a corrosive effect on patients' perceptions of disease, leaving them susceptible to the worst kinds of quackery.

EPILOGUE:
Albert Schweitzer and the Witch Doctor

A Parable

Albert Schweitzer was a musician, philosopher, theologian, and physician. In 1912, using his own money, he established a clinic in Lambaréné, Gabon, in western Africa. Within nine months, more than two thousands natives had come to see him. Schweitzer gave them quinine for malaria, digitalis for heart disease, and salvarsan—the first antibiotic—for syphilis. When patients came to him with strangulated hernias or abdominal tumors, he anesthetized them with chloroform and treated their pain with morphine. Albert Schweitzer brought modern medicine to a small part of Africa.

Toward the end of both of their lives, Norman Cousins, author of *Anatomy of an Illness*, met Albert Schweitzer. "At the dinner table of the Schweitzer Hospital at Lambaréné," wrote Cousins, "I had ventured the remark that local people were lucky to have access to the Schweitzer clinic instead of having to depend on witch-doctor supernaturalism. Dr. Schweitzer asked me how much I knew about witch doctors. I was trapped by my ignorance. The next day the great doctor took me to a nearby jungle clearing where he introduced me to an elderly witch doctor."

"For the next two hours, we stood off to one side and watched," recalled Cousins. "With some patients, the witch doctor merely put herbs in a brown paper bag and instructed the ill person in their use. With other patients, he gave no herbs but filled the air with incantations. A third category of patients he merely spoke to in a subdued voice and pointed to Dr. Schweitzer." On the way back, Schweitzer interpreted what they had seen. The first group of patients had minor illnesses that would resolve on their own or for which modern medicine offered little. The second group had psychological problems treated with "African psychotherapy." The third had massive hernias or extrauterine pregnancies or dislocated shoulders or tumors—diseases the witch doctor couldn't treat—so he directed them to Dr. Schweitzer.

Schweitzer described the value of the witch doctor. "The witch doctor succeeds for the same reason the rest of us succeed," he said. "Each patient carries his own doctor inside him. They come to us not knowing that truth. We are at our best when we give the doctor who resides within each patient a chance to go to work."

In Gabon, both Albert Schweitzer's modern medicine and the witch doctor's ancient medicine had their place. Schweitzer offered specific treatments for treatable diseases, and the witch doctor offered placebo medicine when nothing more was necessary or available. Both recognized the value of the other. Such is the case with today's mainstream and alternative healers: both have their place. The problem comes when mainstream healers dismiss the placebo response as trivial or when alternative healers offer placebos instead of lifesaving medicines or charge an exorbitant price for their remedies or promote therapies as harmless when they're not or encourage magical thinking and scientific denialism at a time when we can least afford it.

As consumers, we have certain responsibilities. If we're going to make decisions about our health, we need to make sure we're not influenced by the wrong things—specifically, that we don't give alternative medicine a free pass because we're fed up with conventional medicine; or buy products because we're seduced by marketing terms such as *natural*, *organic*, and *antioxidant*; or give undeserved credence to celebrities; or make hasty, uneducated decisions because we're desperate to do something, anything, to save ourselves and our children; or fall prey to healers whose charisma obscures the fact that their therapies are bogus. Rather, we need to focus on the quality of scientific studies. And where scientific studies don't exist, we should insist that they be performed. If not, we'll continue to be deceived by therapies whose claims are fanciful.

Making decisions about our health is an awesome responsibility. If we're going to do it, we need to take it seriously. Otherwise we will violate the most basic principle of medicine: First, do no harm.

Acknowledgments

I would like to thank Gail Ross for helping me see the forest through the trees; Gail Winston for her wisdom, humor, medical perspective, and deft editorial hand; Erica DeWald for her research assistance; Peter Adamson, Jeff Bergelson, Sarah Erush, Melissa Ketunuti, Grace Lee, Dan Lee, Ed Milder, Juliette Milder, Phyllis Milder, Charlotte Moser, Bonnie Offit, Emily Offit, Don Mitchell, Anne Reilly, Jason Schwartz, Kirsten Thistle, and Trine Tsouderos for their careful reading of the manuscript; and Peter Adamson, Carol Baker, Stephen Barrett, Art Caplan, Anne Gershon, Jeanine Graf, Peter Barton Hutt, John Maris, Emily Rosa, Linda Rosa, Larry Sarner, Gene Shapiro, and Alison Singer for their expertise.

Notes

PROLOGUE: TAKING A LOOK AT ALTERNATIVE MEDICINE

1   Widespread use of alternative therapies: M. Conley, "Vitamins and Vitamin Supplements: Use Increases in America," ABC News, April 13, 2011; A. Abdel-Rahman, N. Anyangwe, L. Carlacci, et al., "The Safety and Regulation of Natural Products Used as Foods and Food Ingredients," *Toxicological Sciences* 123 (2011): 333–48.

1   Hospital survey of alternative therapies: S. Ananth, "2010 Complementary and Alternative Medicine Survey of Hospitals," Samueli Institute, Alexandria, Va.

2   Pfizer acquires Alacer: "Pfizer Acquires Alacer Corp., a Leading Vitamin Supplement Company," press release, Business Wire, February 27, 2012.

INTRODUCTION: SAVING JOEY HOFBAUER

7   Joey Hofbauer: L. J. Lefkowitz, Attorney General, State of New York, by D. K. McGivney, Esq., Appendix on Appeal, *In the Matter of Joseph Hofbauer,* State of New York Supreme Court, Appellate Division, Third Judicial Department, index no. N-46-1164-77, May 17, 1978.

8   Treatment of Hodgkin's disease: E. C. Easson and M. H. Russell, "Cure of Hodgkin's Disease," *British Medical Journal* 1 (1963): 1704–7; V. T. Devita, A. R. Secpick, and P. P. Carbone, "Combination Chemotherapy in the Treatment of Advanced Hodgkin's Disease," *Annals of Internal Medicine* 73 (1970): 881–95; P. P. Carbone, H. S. Kaplan, K. Musshoff, et al., "Report of the Committee on Hodgkin's Disease Staging Classification," *Cancer*

Research 31 (1971): 1860–61; S. A. Rosenberg, "Development of the Concept of Hodgkin's Disease as a Curable Illness: The American Experience," in P. M. Mauch, J. O. Armitage, V. Diehl, R. T. Hoppe, and L. M. Weiss, eds., *Hodgkin's Disease* (Philadelphia: Lippincott Williams & Wilkins, 1999): 47–57; J. O. Armitage, "Current Concepts: Early Stage Hodgkin's Lymphoma," *New England Journal of Medicine* 363 (2010): 653–62.

8   Hofbauers reject Cohn's advice: Lefkowitz, "Hofbauer," A49–A74; 411 N.Y.S.2d 416; 47 N.Y.2d 648.

9   Chagnon's letter to the Hofbauers: Lefkowitz, "Hofbauer," A1546–A1547.

9   States' rights in protection of a minor: 411 N.Y.S.2d 416.

10   Sheridan regarding visit to the Hofbauers: Lefkowitz, "Hofbauer," A1180–A1181.

10   Joey Hofbauer at St. Peter's: Ibid., A1031.

10   John Hofbauer regarding laetrile at St. Peter's: Ibid., A1043.

11   John Hofbauer searches for doctor: Ibid., A1028.

11   Schachter consent form: Ibid., A1271–A1273.

11   Judge Brown offers six-month extension: 411 N.Y.S.2d 416; 393 N.E.2d 1009.

12   Schachter's therapies: V. Herbert, "Laetrile: The Cult of Cyanide: Promoting Poison for Profit," *The American Journal of Clinical Nutrition* 32 (1979): 1149–51; Lefkowitz, *Hofbauer*, A164.

12   Deaths from coffee enemas: J. W. Eisele and D. T. Reay, "Deaths Related to Coffee Enemas," *Journal of the American Medical Association* 244 (1980): 1608–9.

12   New York State laws on human experimentation: Public Health Law, State of New York, Article 24-A, effective September 1, 1975: Protection of Human Subjects, Sections 2440–46.

12   "Witch doctor's diet": Ibid., A149.

12   Horton testimony: Ibid., A811–A872.

13   Tartaglia testimony: Ibid., A782–A900.

13   Schachter's indiscretions: Herbert, "Laetrile," 1149–51.

13   Schachter regarding Joey's progress: Lefkowitz, "Hofbauer," A925–A972.

13   Laetrile promoters testifying at Hofbauer trial: Herbert, "Laetrile," 1149–51; L. J. Lefkowitz, Attorney General, State of New York, by D. K. McGivney, Esq., Appendix on Appeal, *In the Matter of Joseph Hofbauer,* State of New York Supreme Court, Appellate Division, Third Judicial Department, index no. N-46-1164-77, May 17, 1978.

13   Judge Brown's verdict: 411 N.Y.S.2d 416; 393 N.E.2d 1009.

14 Influence of John Birch Society: J. H. Young, *American Health Quackery* (Princeton, N.J.: Princeton University Press, 1992), 218–28; Robert Johnston, *The Politics of Healing: Histories of Alternative Medicine in Twentieth-Century North America* (New York: Routledge, 2004), 237.

14 States legalize laetrile: I. J. Lerner, "Laetrile: A Lesson in Cancer Quackery," *CA: Cancer Journal for Clinicians* 31 (1981): 91–95.

14 Laetrile a billion-dollar business: V. Herbert, "Laetrile," *New England Journal of Medicine* 307 (1982): 119.

14 McQueen's illness: B. Lerner, *When Illness Goes Public: Celebrity Patients and How We Look at Medicine* (Baltimore: Johns Hopkins University Press, 2006), 141.

15 McQueen given no hope: Ibid., 143.

15 Kelley early career: S. Watson and K. MacKay, "McQueen's Holistic Medicine Man: Claims He Cured His Own Cancer with His Holistic Treatments," *People*, October 20, 1980.

15 McQueen's treatment: Lerner, *When Illness Goes Public*, 147; "McQueen Treatment: Laetrile, Megavitamins, Animal Cells," Associated Press, October 10, 1980; C. Sandford, *McQueen: The Biography* (New York: Taylor Trade Publishing, 2001), 427.

15 Kelley on *Tomorrow*: Sandford, *McQueen*.

16 McQueen on Mexican television: Ibid., 435.

16 First Hofbauer appeal: 411 N.Y.S.2d 416.

16 Ingelfinger regarding laetrile: F. J. Ingelfinger, "Laetrilomania," *New England Journal of Medicine* 296 (1977): 1167–68.

17 FDA approves laetrile study: "FDA OK's Testing Laetrile on Humans," *Boston Globe*, January 4, 1980; "U. S. Test of Laetrile on Humans Backed," *New York Times*, January 4, 1980.

17 Edward Kennedy hearing: Young, *American Health Quackery*, 220; J. H. Young, *The Medical Messiahs: A Social History of Health Quackery in Twentieth-Century America* (Princeton, N.J.: Princeton University Press, 1967), 454–57; Herbert, "Laetrile," 1137; Anonymous: "Lopsided 161–58 Vote Defeats Legalizing Laetrile," *Boston Herald American*, May 19, 1978.

17 Jasen decision: 393 N.E.2d 1009.

18 Schachter regarding Hofbauer's death: W. Waggoner, "Boy, 10, in Laetrile Case Dies," *New York Times*, July 18, 1980.

18 Coleman visits McQueen: Sandford, *McQueen*, 438.

18 McQueen dies following surgery: Ibid., 442–43.

18   Moertel study: C. G. Moertel, T. R. Fleming, J. Rubin, et al., "A Clinical Trial of Amygdalin (Laetrile) in the Treatment of Human Cancer," *New England Journal of Medicine* 306 (1982): 201–06.

19   FDA ban on laetrile: Young, *American Health Quackery*, 226.

20   Judge Brown's Finding of Facts: Lefkowitz, "Hofbauer," A0.17.

21   Kirkpatrick Dilling and the NHF: Ibid., A184.

21   Schachter brochure: Schachter Center for Complementary and Alternative Medicine, Two Executive Boulevard, Suite 202, Suffern, NY, obtained October 2009.

Chapter 1: Rediscovering the Past

25   Oprah Winfrey impact: W. Kosova, "Live Your Best Life Ever!" *Newsweek*, May 30, 2009; D. Gorski, "The Oprah-fication of Medicine," *Science-Based Medicine* blog, www.sciencebasedmedicine.org/?p=497.

26   Whitworth and Oz: H. Dreher, "Medicine Goes Mental," *New York*, May 11, 1998, http://nymag.com/nymetro/health/features/2664/#ixzz0fds1lbrm.

29   Bloodletting: R. Shapiro, *Suckers: How Alternative Medicine Makes Fools of Us All* (London: Harvill Secker, 2008), 10–11; C. Wanjek, *Bad Medicine: Misconceptions and Misuses Revealed, from Distance Healing to Vitamin O* (Hoboken, N.J.: John Wiley & Sons, 2003), 7–10; Bob McCoy, *Quack! Tales of Medical Fraud from the Museum of Questionable Medical Devices* (Santa Monica, Calif.: Santa Monica Press, 2000), 25–27; S. Singh and E. Ernst, *Trick or Treatment: The Undeniable Facts About Alternative Medicine* (New York: W. W. Norton, 2008), 7–14; D. Morens, "Death of a President," *New England Journal of Medicine* 341 (1999): 1845–49.

30   Osler regarding bloodletting: J. H. Young, *The Toadstool Millionaires: A Social History of Patent Medicines in America Before Federal Regulation* (Princeton, N.J.: Princeton University Press, 1972), 4.

31   James Lind and scurvy: Singh and Ernst, *Trick or Treatment*, 14–24.

32   Longevity data: R. Tallis, *Hippocratic Oaths: Medicine and Its Discontents* (London: Atlantic Books, 2005), 22; National Center for Health Statistics, *Health, United States, 2010: With Special Feature on Death and Dying* (Hyattsville, Md.: CDC, National Center for Health Statistics, 2011); available at www.cdc.gov/nchs.

32   Dr. Nemeh on Oz show: "Is This Man a Faith Healer?" *The Dr. Oz Show*, www.doctoroz.com/videos/man-faith-healer-pt-1.

33   Pamela Miles: P. Miles, *Reiki: A Comprehensive Guide* (New York: Penguin, 2006), 32.

34   Emily Rosa and therapeutic touch: L. Rosa, E. Rosa, L. Sarner, and S. Barrett, "A Closer Look at Therapeutic Touch," *Journal of the American Medical Association* 279 (1998): 1005–10; G. Kolata, "A Child's Paper Poses a Medical Challenge," *New York Times*, April 1, 1998; M. D. Lemonick, "Emily's Little Experiment," *Time*, April 13, 1998; D. Krieger, *Accepting Your Power to Heal* (Rochester, Vt.: Bear & Company Publishing, 1993); Emily Rosa, Linda Rosa, and Larry Sarner, interviewed by the author, September 23, 2011.

35   John Edward on Oz show: "Are Psychics the New Therapists?" *The Dr. Oz Show*, www.doctoroz.com/videos/are-psychics-new-therapists-pt-1.

36   Edward's psychic strategies: Orac, "When Faith Healing Isn't Woo Enough for Dr. Oz," *ScienceBlogs*, http://scienceblogs.com/insolence/2011/03/when_faith_healing_isnt_enough_woo_for_d.php.

36   Oz regarding Edward's powers: "Are Psychics the New Therapists?" *The Dr. Oz Show*.

37   Andrew Weil background: D. Hurley, *Natural Causes: Death, Lies, and Politics in America's Vitamin and Herbal Supplement Industry* (New York: Broadway Books, 2006), 236–40; M. Specter, *Denialism: How Irrational Thinking Hinders Scientific Progress, Harms the Planet, and Threatens Our Lives* (New York: Penguin Press, 2009), 149–51.

38   Deepak Chopra: S. Barrett and W. T. Jarvis, eds., *The Health Robbers: A Close Look at Quackery in America* (Buffalo: Prometheus Books, 1993), 243–45; Singh and Ernst, *Trick or Treatment*, 256.

39   Homeopathy: L. Silver, *Challenging Nature: The Clash Between Biotechnology and Spirituality* (New York: Ecco, 2006), 250–53.

39   Chiropractic: Singh and Ernst, *Trick or Treatment*, 156–66.

40   Efficacy of chiropractic manipulations: Ibid., 166–67.

41   Weil regarding time with patients: "Dr. Andrew Weil: The Future of Medicine," *The Dr. Oz Show*, www.doctoroz.com/videos/dr-andrew-weil-future-medicine-pt-1.

41   Oz regarding customized therapy: "Controversial Medicine: Alternative Health," *The Dr. Oz Show*, www.doctoroz.com/videos/alternative-medicine-controversy-pt-1.

41   Oz regarding Chinese medicine: Ibid.

41   Scientist from *2012*: *2012*, directed by Roland Emmerich (2009; Culver City, Calif.: Sony Pictures Digital, 2010).

42   Tallis regarding ancient wisdom: Tallis, *Hippocratic Oaths*, 29.

42   Health care choices in China, Hong Kong, and Japan: J. Diamond, *Snake Oil and Other Preoccupations* (London: Vintage, 2001), 27.

42   Acupuncture in China: R. Slack, "Acupuncture: A Science-Based Assessment," position paper, Center for Inquiry, 2010.

42   Diamond regarding medicine in Africa: Diamond, *Snake Oil*, 26–27.

43   Weinberg regarding the universe: R. Park, *Superstition: Belief in the Age of Science* (Princeton, N.J.: Princeton University Press, 2008), 5.

43   Oz regarding empowerment of alternative medicine: "Controversial Medicine: Alternative Health," *The Dr. Oz Show*, www.doctoroz.com/videos/alternative-medicine-controversy-pt-1.

CHAPTER 2: THE VITAMIN CRAZE

48   Gann quote: C. Gann, "Should You Take a Multivitamin?" ABC News, October 12, 2011.

49   Pauling's early life: T. Goertzel and B. Goertzel, *Linus Pauling: A Life in Science and Politics* (New York: Basic Books, 1995), 1, 35; D. Newton, *Linus Pauling: Scientist and Advocate* (New York: Facts on File, 1994), 20, 35; T. Hager, *Force of Nature: The Life of Linus Pauling* (New York: Simon & Schuster, 1995), 50; C. Mead and T. Hager, *Linus Pauling: Scientist and Peacemaker* (Corvallis: Oregon State University Press, 2001), 79.

49   Pauling's work on chemical bonding: Goertzel and Goertzel, *Linus Pauling*, 77; Newton, *Linus Pauling*, 30–38; Hager, *Force of Nature*, 52–60, 157–62; B. Marinacci, ed., *Linus Pauling in His Own Words: Selections from His Writings, Speeches, and Interviews* (New York: Simon & Schuster, 1995), 79–88.

49   Pauling's work on sickle-cell anemia: Hager, *Force of Nature*, 332–34; Hager, *Linus Pauling*, 87–89; Newton, *Linus Pauling*, 69; Goertzel and Goertzel, *Linus Pauling*, 90.

50   Pauling and the alpha helix: Goertzel and Goertzel, *Linus Pauling*, 91–94; Mead and Hager, *Linus Pauling*, 14.

50   Pauling and evolutionary biology: Hager, *Force of Nature*, 540–46; Mead and Hager, *Linus Pauling*, 169–76; Hager, *Linus Pauling*, 119–20; M. F. Perutz, "Linus Pauling: 1901–1994," *Structural Biology* 1 (1994): 667–71.

50   Pauling's activism for world peace: Mead and Hager, *Linus Pauling*, 13–17; Serafini, *Linus Pauling*, 186–90; Newton, *Linus Pauling*, 59–83; Goertzel and Goertzel, *Linus Pauling*, 143–47; Marinacci, *Linus Pauling*, 184.

51   Pauling's awards and honors: Serafini, *Linus Pauling*, xxii; Mead and Hager, *Linus Pauling*, 16, 18; Newton, *Linus Pauling*, 69, 109; Goertzel and Goertzel, *Linus Pauling*, 111.

51   Pauling as "classic tragedy": Goertzel and Goertzel, *Linus Pauling*, xvi.

51   Pauling meets Stone: Marinacci, *Linus Pauling*, 246.

52   Stone's credentials: Barrett and Jarvis, *Health Robbers*, 386.

52   Linus Pauling takes vitamin C: Marinacci, *Linus Pauling*, 246.

52   Pauling predicts end of the common cold: L. Pauling, *Vitamin C and the Common Cold* (San Francisco: W. H. Freeman and Company, 1970), 6.

52   Impact of *Vitamin C and the Common Cold*: Hager, *Linus Pauling*, 126; Hager, *Force of Nature*, 583.

53   Minnesota study: Goertzel and Goertzel, *Linus Pauling*, 203.

53   Maryland study: Hager, *Force of Nature*, 582.

53   Toronto study: S. Barrett, *Health Schemes, Scams, and Frauds* (Mount Vernon, N.Y.: Consumer Reports Books, 1990), 57.

53   Netherlands study: Hurley, *Natural Causes*, 172.

53   Fifteen studies of vitamin C don't support Pauling's claims: S. Barrett, W. London, R. Bartz, and M. Kroger, eds., *Consumer Health: A Guide to Intelligent Decisions* (New York: McGraw-Hill, 2007), 246.

53   Public health associations don't support Pauling's claims: Hager, *Linus Pauling*, 126; Goertzel and Goertzel, *Linus Pauling*, 203; Hurley, *Natural Causes*, 165–66; Hager, *Force of Nature*, 578.

54   Pauling and allergies: Goertzel and Goertzel, *Linus Pauling*, 201.

54   Ewan Cameron and Glasgow study: Hager, *Force of Nature*, 583–86; Newton, *Linus Pauling*, 105–6.

54   Cameron and Pauling study rejected by academy journal: Hager, *Force of Nature*, 585–86.

54   Criticism of Cameron and Pauling study: Barrett and Jarvis, *Health Robbers*, 386.

55   Pauling claims 75 percent success rate: Newton, *Linus Pauling*, 106.

55   Pauling claims Americans will live longer: Pauling, *Live Longer*, 243.

55   Maris and Nobel Prize: John Maris, interviewed by the author, September 28, 2011.

55   Moertel's first study: E. T. Creagan, C. G. Moertel, J. R. O'Fallon, et al., "Failure of High-Dose Vitamin C (Ascorbic Acid) Therapy to Benefit Patients with Advanced Cancer: A Controlled Trial," *New England Journal of Medicine* 301 (1979): 687–90.

55   Moertel bullied to perform second study: Goertzel and Goertzel, *Linus Pauling*, 215.

56   Moertel's second study: C. G. Moertel, T. R. Fleming, E. T. Creagon, et al., "High-Dose Vitamin C Versus Placebo in the Treatment of Patients with Advanced Cancer Who Have Had No Prior Chemotherapy: A Randomized Double-Blind Comparison," *New England Journal of Medicine* 312 (1985): 137–41.

56   Cameron regarding Pauling's anger: Goertzel and Goertzel, *Linus Pauling*, 216–17.

56   Pauling attacks Moertel's second study: Ibid.

56   Pauling considers suing Moertel: Ibid.

56   Continued lack of evidence supporting vitamin C as a cure for cancer: G. van Poppel and H. van den Berg, "Vitamins and Cancer," *Cancer Letters* 114 (1997): 195–202; S. J. Padayatty, A. Katz, Y. Wang, et al., "Vitamin C as an Antioxidant: Evaluation of Its Role in Disease Prevention," *Journal of the American College of Nutrition* 22 (2003): 18–35.

56   Pauling recommends high doses of several vitamins: Barrett et al., *Consumer Health*, 246.

56   Pauling claims vitamins as cure-all: Goertzel and Goertzel, *Linus Pauling*, 204; I. Stone, *The Healing Factor: Vitamin C Against Disease* (New York: Grosset & Dunlap, 1972), ix-x.

57   Pauling claims vitamin C treats AIDS: Young, *American Health Quackery*, 260.

57   *Time* article: A. Toufexis, J. M. Horowitz, E. Lafferty, and D. Thompson, "The New Scoop on Vitamins," *Time*, April 6, 1992.

57   NNFA regarding impact of *Time* article: Barrett, *Vitamin Pushers*, 369–70.

57   Toufexis regarding impact of *Time* article: Ibid., 370.

58   Finnish antioxidant study: Alpha-Tocopherol, Beta-Carotene Cancer Prevention Study Group, "The Effect of Vitamin E and Beta-Carotene on the Incidence of Lung Cancer and Other Cancers in Male Smokers," *New England Journal of Medicine* 330 (1994): 1029–35.

59   Asbestos exposure antioxidant study: G. E. Goodman, M. D. Thornquist, J. Balmes, et al., "The Beta-Carotene and Retinol Efficacy Trial," *Journal of the National Cancer Institute* 96 (2004): 1743–50.

59   2004 antioxidant study: G. Bjelakovic, D. Nikolova, R. G. Simonetti, et al., "Antioxidant Supplements for Prevention of Gastrointestinal Cancers: A Systematic Review and Meta-Analysis," *Lancet* 364 (2005): 37–46 (italics mine).

59   Review of nineteen antioxidant studies: E. R. Miller III, R. Pastor-Barriuso, D. Dalal, et al., "Meta-Analysis: High-Dosage Vitamin E Supplementation May Increase All-Cause Mortality," *Annals of Internal Medicine* 142 (2005): 37–46.

59   Caballero on antioxidant studies: G. Kolata, "Large Doses of Vitamin E May Be Harmful," *New York Times*, November 11, 2004.

60   *JAMA* antioxidant study: E. Lonn, J. Bosch, S. Yusuf, et al., "Effects of Long-Term Vitamin E Supplementation on Cardiovascular Events and Cancer," *Journal of the American Medical Association* 293 (2005): 1338–47.

60   Antioxidant study of prostate cancer: K. A. Lawson, M. E. Wright, A. Subar, et al., "Multivitamin Use and Risk of Prostate Cancer in the National Institutes of Health—AARP Diet and Health Study," *Journal of the National Cancer Institute* 99 (2007) 754–64.

60   Cochrane review of antioxidant studies: G. Bjelakovic, D. Nikolova, L. L. Gluud, et al., "Antioxidant Supplements for Prevention of Mortality in Healthy Participants and Patients with Various Diseases," Cochrane Database of Systematic Reviews 3 (2012): 2. Art. no. CD007176, DOI: 10.1002/14651858.CD007176.pub2.

60   Minnesota supplement study: J. Mursu, K. Robien, L. J. Harnack, et al., "Dietary Supplements and Mortality Rate in Older Women: The Iowa Women's Health Study," *Archives of Internal Medicine* 171 (2011): 1625–33; E. Brown, "Dietary Supplements Risk for Older Women, Study Finds," *Los Angeles Times*, October 10, 2011; J. C. Dooren, "Supplements Offer Risks, Little Benefit, Study Says," *Wall Street Journal*, October 11, 2011.

60   Cleveland Clinic Study and *Wall Street Journal* headline: E. A. Klein, I. M. Thompson Jr., C. M. Tangen, et al., "Vitamin E and the Risk of Prostate Cancer: The Selenium and Vitamin E Cancer Prevention Trial (SELECT)," *Journal of the American Medical Association* 306 (2011): 1549–56; S. S. Wang, "Is This the End of Popping Vitamins?" *Wall Street Journal*, October 25, 2011.

61   Fortunato quote: S. S. Wang, "Is This the End of Popping Vitamins?" *Wall Street Journal*, October 25, 2011.

61   "The antioxidant paradox": B. Halliwell, "The Antioxidant Paradox," *Lancet* 355 (2000): 1179–80.

61   Pauling's interview at Oregon State University: Serafini, *Linus Pauling*, 263.

65   Medical huckster: A. Anderson, *Snake Oil, Hustlers and Hambones: The American Medicine Show* (Jefferson, N.C.: McFarland & Company, 2000), 1.

66   List of patent medicines: Hurley, *Natural Causes*, 30–31.

66     $75 million business: Ibid.

66   Harvey Wiley: Young, *Toadstool Millionaires*, 226–44.

67   Samuel Adams and "The Great American Fraud": Young, *Toadstool*, 205–25.

67   Coca-Cola: J. Schwarcz, *Science, Sense and Nonsense: 61 Nourishing, Healthy, Bunk-Free Commentaries on the Chemistry That Affects Us All* (Scarborough, Ontario: Doubleday Canada, 2009), 193.

68   Wiley's proposed legislation: Ibid., 226.

68   Proprietary Association of America: Ibid.

68   Upton Sinclair and *The Jungle*: U. Sinclair, *The Jungle* (New York: Bantam Books, 1906).

69   The Pure Food and Drug Act of 1906: Young, *Toadstool*, 226–44.

70   Elixir Sulfanilamide: E. M. K. Geiling, and P. R. Cannon, "Pathologic Effects of Elixir of Sulfanilamide (Diethylene Glycol) Poisoning," *Journal of the American Medical Association* 111 (1938): 919–26; P. N. Leech, "Elixir of Sulfanilamide-Massengill," *Journal of the American Medical Association* 109 (1937): 1531–39; R. Steinbrook, "Testing Medications in Children," *New England Journal of Medicine* 347 (2002): 1462–70; P. M. Wax, "Elixirs, Diluents, and the Passage of the 1938 Federal Food, Drug, and Cosmetic Act," *Annals of Internal Medicine* 122 (1995): 456–61; "Fatal Elixir Seized as Adulterated," *New York Times*, October 30, 1937; " 'Death Drug' Hunt Covered 15 States," *New York Times*, November 26, 1937.

71   The Food, Drug, and Cosmetics Act of 1938: Barrett et al., *Consumer Health*, 535–37; Anderson, *Snake Oil*, 158–59; Hurley, *Natural Causes*, 35; Barrett and Herbert, *Vitamin Pushers*, 74; Young, *Health Quackery*, 269; V. Herbert, and S. Barrett, *Vitamins and "Health" Foods: The Great American Hustle* (Philadelphia: George F. Stickley Company, 1981), 149.

71   Thalidomide disaster: Barrett et al., *Consumer Health*, 535–37; Singh and Ernst, *Trick or Treatment*, 273.

72   Kefauver-Harris Amendment: Barrett et al., *Consumer Health*, 535–37; Hurley, *Natural Causes*, 36.

73   FDA's attempt to regulate vitamins: Ibid., 47–53.

73   NHF founders and agenda: Barrett and Jarvis, *Health Robbers*, 398–408; E. Juhne, *Quacks and Crusaders: The Fabulous Careers of John Brinkley, Norman Baker, & Harry Hoxsey* (Lawrence: University Press of Kansas, 2002), 150.

75   Proxmire Amendment: Hurley, *Natural Causes*, 47–53.

77   Hutt regarding FDA's mistake: P. B. Hutt, "U. S. Government Regulation of Food with Claims for Special Physiological Value," in M. K. Schmidt and T. P. Lapuza, *Essentials of Functional Foods* (New York: Springer, 2000).

77   Hurley regarding Proxmire Amendment: Hurley, *Natural Causes*, 53.

78   Kessler regarding unsubstantiated claims: Ibid., 92.

78   Food, Drug, Cosmetic, and Device Enforcement Amendments: Ibid., 78.

79   Hutt regarding Kessler: Peter Barton Hutt, interviewed by the author, October 10, 2011.

80   Gerry Kessler vs. the Food, Drug, Cosmetic, and Device Enforcement Amendments: Hurley, *Natural Causes*, 72–103.

80   Orrin Hatch and ties to the supplement industry: Ibid., 75–77.

83   Gerry Kessler and the Dietary Supplement Health and Education Act: Ibid., 72–103.

87   The Snake Oil Protection Act: "The 1993 Snake Oil Protection Act," *New York Times*, October 5, 1993.

87   Vioxx study in 2000: C. Bombardier, L. Laine, A. Reicin, et al., "Comparison of Upper Gastrointestinal Toxicity of Rofecoxib and Naproxen in Patients with Rheumatoid Arthritis," *New England Journal of Medicine* 343 (2000): 1520–28.

88   Vioxx study in 2005: R. S. Bresalier, R. S. Sandler, H. Quan, et al., "Cardiovascular Events Associated with Rofecoxib in a Colorectal Adenoma Chemoprevention Trial," *New England Journal of Medicine* 352 (2005): 1092–1102.

88   Gilmartin regarding Vioxx withdrawal: J. Kelly, "Vioxx Hearing Raises Questions About What Merck Knew and When," Medscape, www.medscape.com/viewarticle/538025.

90   Toxic products contained in foods: Food Protection Committee, Food and Nutrition Board, National Academy of Sciences, National Research Council. *Toxicants Occurring Naturally in Foods*. Washington, D. C.: National Academy of Sciences Press, 1966.

90   Singh and Ernst regarding natural vs. unnatural products: Singh and Ernst, *Trick or Treatment*, 222.

90   Harm from herbal products: A. J. Tomassoni and K. Simone, "Herbal Medicines for Children: An Illusion of Safety?" *Current Opinion in Pediatrics* 13 (2001): 162–69; A. D. Wolff, "Herbal Remedies and Children: Do They Work? Are They Harmful?" *Pediatrics* 112 (2003): 240–46; I. Choonara, "Safety of Herbal Medicines in Children," *Archives of Diseases of Children* 88 (2003): 1032–33; Hurley, *Natural Causes*, 142–43; A. Abdel-Rahman, N. Anyangwe, L. Carlacci, et al., "The Safety and Regulation of Natural Products Used as Foods and Food Ingredients," *Toxicological Sciences* 123 (2011): 333–48.

90   Selenium disaster: T. Tsouderos, "Dietary Supplements: Manufacturing Troubles Widespread, FDA Inspections Show," *Chicago Tribune*, June 30, 2012.

91   Deaths from herbal products: A. J. Tomassoni and K. Simone, "Herbal Medicines for Children: An Illusion of Safety?" *Current Opinion in Pediatrics* 13 (2001): 162–69.

91   OxyElite Pro: V. Taylor, "Supplement OxyElite Pro Linked to Liver Failure, Hepatitis," *New York Daily News*, October 9, 2013.

91   Lack of safety testing for dietary supplements: G. Lundberg, "The Wild World of American 'Supplements,' " *MedPage Today*, March 5, 2012.

91   Harvard study of ayurvedic remedies: R. B. Saper, S. N. Kales, J. Paquin, et al., "Heavy Metal Content of Ayurvedic Herbal Medicine Products," *Journal of the American Medical Association* 292 (2005): 2868–73.

91   Deaths from ayurvedic remedies: Ibid.; J. Kew, C. Morris, A. Aihie, et al., "Arsenic and Mercury Intoxication Due to Indian Ethnic Remedies," *British Medical Journal* 306 (1993): 506–507; C. Moore and R. Adler, "Herbal Vitamins: Lead Toxicity and Developmental Delay," *Pediatrics* 106 (2000): 600–602; Centers for Disease Control and Prevention, "Lead Poisoning Associated with Ayurvedic Medications—Five States, 2000–2003," *Morbidity and Mortality Weekly Report* 53 (2004): 582–84.

91   Antimony contamination: T. Tsouderos, "Dietary Supplements."

91   Anabolic steroids: D. Ricks, "Recall of Purity First Vitamins Widens," *Newsday*, August 2, 2013.

91   FDA estimates of adverse events caused by dietary supplements: Hurley, *Natural Causes*, 164; D. M. Marcus and A. P. Grollman, "The Consequences of Ineffective Regulation of Dietary Supplements," *Archives of Internal Medicine* 172 (2012): 1035–36.

91   Purity First contamination: D. Ricks, "Recall of Purity First Vitamins Widens."

91   OxyElite Pro: Food and Drug Administration Consumer Updates: "OxyElite Pro Supplements Recalled, http://www.fda.gov/forconsumers/consumerupdates/ucm374742.htm. Accessed November 17, 2013.

91   Herbal products contain other herbs: S. G. Newmaster, M. Grguric, D. Shanmughanandhan, et al., "DNA Barcoding Detects Contamination and Substitution in North American Herbal Products, "*BMC Medicine* 11 (2013): 222–234.

92   Harris poll: Hurley, *Natural Causes*, 19.

93   FDA inspects supplement makers: T. Tsouderos, "Dietary Supplements."

93   Sales of dietary supplements: D. Marcus, "Consumer Reports and Alternative Therapies," *Science-Based Medicine* blog, http://sciencebasedmedicine.org/index.php/consumer-reports-and-alternative-therapies.

CHAPTER 4: FIFTY-FOUR THOUSAND SUPPLEMENTS

95   Herbal remedies: "The Alternative Health Controversy," *The Dr. Oz Show*, www.doctoroz.com/videos/alternative-health-controversy-pt-1.

95   Rotavirus vaccines: T. Vesikari, D. O. Matson, P. Dennehy, et al., "Safety and Efficacy of a Pentavalent Human-Bovine (WC3) Reassortant Rotavirus Vaccine," *New England Journal of Medicine* 354 (2006): 23–33; G. M. Ruiz-Palacios, I. Perez-Schael, F. R. Veláquez, et al., "Safety and Efficacy of an Attenuated Vaccine Against Severe Rotavirus Gastroenteritis," *New England Journal of Medicine* 354 (2006): 11–21.

96   NCCAM studies: T. Tsouderos, "Bad Science, Suspect Medicine," *Chicago Tribune*, December 11, 2011; E. V. Mielczarek and B. D. Engler, "Measuring Mythology: Startling Concepts in NCCAM Grants," *Skeptical Inquirer* 36 (2012): 35–43.

97   Medawar quote: Singh and Ernst, *Trick or Treatment*, 87 (italics mine).

98   Novella regarding herbal remedies: Steven Novella, "A Skeptic in Oz," *Neurologica* blog, http://theness.com/neurologicablog/index.php/a-skeptics-in-oz.

99   Ginkgo and dementia: S. T. DeKosky, J. D. Williamson, A. L. Fitzpatrick, et al., "Ginkgo Biloba for Prevention of Dementia: A Randomized Controlled Trial," *Journal of the American Medical Association* 300 (2008): 2253–62; B. Vellas, N. Coley, P.-J. Ousset, et al., "Long-Term Use of Standardised Ginkgo Biloba for the Prevention of Alzheimer's Disease

(GuidAge): A Randomised Placebo-Controlled Trial," *Lancet Neurology* 11 (2012): 851–59.

99   St. John's wort and depression: Hypericum Depression Trial Study Group, "Effect of Hypericum Perforatum (St. John's Wort) in Major Depressive Disorder: A Randomized, Controlled Trial," *Journal of the American Medical Association* 287 (2002): 1807–14.

100   Garlic and cholesterol: C. D. Gardner, L. D. Lawson, E. Block, et al., "Effect of Raw Garlic Vs. Commercial Garlic Supplements on Plasma Lipid Concentrations in Adults with Moderate Hypercholesterolemia," *Archives of Internal Medicine* 167 (2007): 346–53.

101   Saw palmetto and prostate enlargement: S. Bent, C. Kane, K. Shinohara, et al., "Saw Palmetto for Benign Prostatic Hyperplasia," *New England Journal of Medicine* 354 (2006): 557–66.

101   Second study of saw palmetto: K. J. Kreder, A. L. Avins, D. Nickel, et al., "Effect of Increasing Doses of Saw Palmetto Extract on Lower Urinary Tract Symptoms," *Journal of the American Medical Association* 306 (2011): 1344–51; D. W. Freeman, "Saw Palmetto No Help for Enlarged Prostate, Study Says," CBS News, September 28, 2011.

102   Milk thistle for hepatitis: M. Freed, "A Randomized, Placebo-Controlled Trial of Oral Silymarin (Milk Thistle) for Chronic Hepatitis C: Final Results of the SYNCH Multicenter Study," presented November 8, 2011, at the American Association for the Study of Liver Diseases annual meeting in San Francisco.

102   Chondroitin sulfate and glucosamine for arthritis: D. O. Clegg, D. J. Reda, C. I. Harris, et al., "Glucosamine, Chondroitin Sulfate, and the Two in Combination for Painful Knee Osteoarthritis," *New England Journal of Medicine* 354 (2006): 795–808.

103   Echinacea for colds: J. A. Taylor, W. Weber, L. Standish, et al., "Efficacy and Safety of Echinacea in Treating Upper Respiratory Tract Infections in Children," *Journal of the American Medical Association* 290 (2003): 2824–30.

103   Benefits of supplements: Hurley, *Natural Causes*, 260–61.

104   Omega-3 fatty acids: P. M. Kris-Etherton, W. S. Harris, L. J. Appel for the Nutrition Committee, "Fish Consumption, Fish Oil, Omega-3 Fatty Acids, and Cardiovascular Disease," *Circulation* 106 (2002): 2747–57; E. C. Rizos, E. E. Nitzani, E. Bika, et al., "Association Between Omega-3 Fatty Acid Supplementation and Risk of Major Cardiovascular Disease Events: A Systematic Review and Meta-Analysis," *Journal of the American Medical Association* 308 (2012): 1024–33; The Risk and Study Prevention

Collaborative Group, "n-3 Fatty Acids in Patients with Multiple Cardio-vascular Risk Factors," *New England Journal of Medicine* 368 (2013): 1800–1808; T. M. Brasky, A. K. Darlee, X. Song, et. al., "Plasma Phospholipid Fatty Acids and Prostate Cancer: Risk in the SELECT Trial," *Journal of the National Cancer Institute* 105 (2013), 1132–1141.

104 Calcium: "Calcium Supplements: Risks and Benefits," Medscape, www.medscape.com/viewarticle/497826_print; "Dietary Supplement Fact Sheet: Calcium," National Institutes of Health, http://ods.od.nih.gov/factsheets/calcium-healthprofessional; Mayo Clinic Staff, "Calcium and Calcium Supplements: Achieving the Right Balance," Mayo Clinic, http://www.mayoclinic.com/health/calcium-supplements/MY01540; G. Kolata, "Healthy Women Advised Not to Take Calcium and Vitamin D to Prevent Fractures," *New York Times*, June 12, 2012.

105 Vitamin D: "Dietary Supplement Fact Sheet: Vitamin D," National Institutes of Health, http://ods.od.nih.gov/factsheets/VitaminD-QuickFacts; "Vitamin D," Mayo Clinic, http://www.mayoclinic.com/health/vitamin-d/NS_patient-vitamind; "Find a Vitamin or Supplement: Vitamin D," WebMD; "What Is Vitamin D? What Are the Benefits of Vitamin D?" *Medical News Today*; Kolata, "Healthy Women Advised"; H. A. Bischoff-Ferrari, W. C. Willett, E. J. Oray, et al., "A Pooled Analysis of Vitamin D Dose Requirements for Fracture Prevention," *New England Journal of Medicine* 367 (2012): 40–49; R. P. Heaney, "Vitamin D—Baseline Status and Effective Dose," *New England Journal of Medicine* 367 (2012): 77–78.

106 Folic acid: "Dietary Supplement Fact Sheet: Folate," http://ods.od.nig.gov/factsheets/Folate-HealthProfessional; "Find a Vitamin or Supplement: Folic Acid," WebMD; "Folic Acid," www.womenshealth.gov; Centers for Disease Control and Prevention, "Folic Acid Recommendations," www.cdc.gov/ncbddd/folicacid/recommendations/html.

107 Multivitamin study: S. P. Fortman, B. U. Burda, C. A. Senger, et al., "Vitamin and Mineral Supplements in the Primary Prevention of Cardiovascular Disease and Cancer: An Updated Systematic Evidence Review for the U.S. Preventative Services Task Force," *Annals of Internal Medicine*, November 12, 2013.

107 Heart specialist regarding vitamins in food: M. Fox, "Vitamins Don't Prevent Heart Disease or Cancer, Experts Find," NBC News, November 11, 2013.

107 Harvard Nutrition Website: Harvard Health Publications, "Getting Your Vitamins and Minerals Through Diet," http://www.health.harvard.edu/

newsletters/Harvard_Womens_Health_Watch/2009/July?Gettin-your
-vitamins-and-minerals-through-diet?print=1.

Chapter 5: Menopause and Aging

111   Celebrities offer medical advice: John Stossel, Fox Business Network, February 24, 2011; Singh and Ernst, *Trick or Treatment*, 251.

113   Somers on Larry King Live: *Larry King Live*, October 14, 2006.

113   Somers and Mack truck: *The Oprah Winfrey Show*, January 29, 2009.

113   Christiane Northrup on menopause: *The Oprah Winfrey Show*, January 15, 2009.

113   The Seven Dwarfs of Menopause: S. Somers, *The Sexy Years: Discover the Hormone Connection: The Secret to Fabulous Sex, Great Health, and Vitality, for Women and Men* (New York: Three Rivers Press, 2004), 2.

114   Somers finds the answer to menopause: S. Somers, *Ageless: The Naked Truth About Bioidentical Hormones* (New York: Three Rivers Press, 2006), 5.

114   O'Donnell on CNN: *The Joy Behar Show*, February 17, 2011.

115   Oprah and bioidentical hormones: W. Kosova, "Live Your Best Life Ever!" *Newsweek*, May 30, 2009.

115   2002 Women's Health Initiative study: J. E. Rossouw, G. L. Anderson, R. L. Prentice, et al., "Risks and Benefits of Estrogen Plus Progestin in Healthy Postmenopausal Women: Principal Results from the Women's Health Initiative Randomized Controlled Trial," *Journal of the American Medical Association* 288 (2002): 321–33.

115   Somers sends the dwarfs packing: Somers, *Sexy Years*, 4.

116   Streicher regarding chemical structures: *The Oprah Winfrey Show*, January 29, 2009.

116   Schwarcz regarding natural substances: J. Schwarcz, *The Fly in the Ointment: 70 Fascinating Commentaries on the Science of Everyday Food & Life* (Toronto: ECW Press, 2004), 129.

116   Streicher regarding hormone manufacture: *The Oprah Winfrey Show*, January 29, 2009.

117   Utian and the Tooth Fairy: *The Oprah Winfrey Show*, January 15, 2009.

117   Streicher and dangerous perceptions: *The Oprah Winfrey Show*, January 29, 2009.

117   Unreliability of compounding pharmacies: American College of Obstetrics and Gynecology, "No Scientific Evidence Supporting Effectiveness

or Safety of Compounded Bioidentical Hormone Therapy," press release, October 31, 2005.

118 Somers regarding hormones and aging: Somers, *Ageless*, xxiii.

118 Somers on hormones as anti-aging medicine: Ibid., 12–14.

118 Somers's anti-aging regimen: Somers, *Sexy Years*, 297; Somers, *Ageless*, 317; Somers, *Sexy Forever*, 130, 139; Somers, *Breakthrough: Eight Steps to Wellness: Life-Altering Secrets from Today's Cutting-Edge Doctors* (New York: Three Rivers Press, 2008), 66–70; *The Oprah Winfrey Show*, January 29, 2009.

119 Oprah and "quackadoo": W. Kosova, "Live Your Best Life Ever!"

119 Somers on reinvigorated sex life: Somers, *Ageless*, 35.

119 Simon Cowell and anti-aging vitamins: A. Jha, "Scientists Buzz Simon Cowell for Promoting Pseudoscience," *The Guardian*, December 27, 2011.

120 Statement by experts on aging: S. J. Oshansky, L. Hayflick, and B. A. Carnes, "No Truth to the Fountain of Youth," *Scientific American*, www.scientificamerican.com/article.cfm?id=no-truth-to-the-fountain-of -youth&print=true.

120 Somers regarding conspiracy: Somers, *Ageless*, 27.

121 Northrup on pharmaceutical companies: C. Northrup, *Women's Bodies, Women's Wisdom: Creating Physical and Emotional Health and Healing* (New York: Bantam Books, 2010), 546–47.

121 Anti-aging industry: S. Jacoby, *Never Say Die: The Myth and Marketing of the New Old Age* (New York: Pantheon Books, 2011), 91.

122 Experts on aging regarding oxidation: Oshansky, Hayflick, and Carnes, "No Truth to the Fountain of Youth."

123 Experts on likely effect of antioxidants: Ibid.

123 Experts on harm of anti-aging drugs: Ibid.

124 Somers on *Larry King Live*: *Larry King Live*, October 14, 2006.

125 Somers on Botox: Somers, *Ageless*, 381.

125 Somers and the FaceMaster: *Larry King Live*, October 14, 2006.

125 Somers's stem-cell face-lift: M. Strobel, "Suzanne Somers: A Supermarket Scare," *Toronto Sun*, February 3, 2011; "Suzanne Somers Plastic Surgery Disaster Shocker," *National Enquirer*, February 2, 2011, www.national enquirer.com/print/33469.

126 Somers's stem-cell breast reconstruction: L. Hamm, "Suzanne Somers Gets Experimental Breast Reconstruction," *People*, February 4, 2012, www.people.com/people/article/0,,20567432,00.html.

126 Jacoby on aging: Jacoby, *Never Say Die*, xi, 5.

126   Somers in 2041: Somers, *Breakthrough*, 1.

127   Jacoby on gullible public: Jacoby, *Never Say Die*, 90–91.

CHAPTER 6: AUTISM'S PIED PIPER

129   Oprah praises McCarthy: *The Oprah Winfrey Show*, September 24, 2008.

129   Rimland paper: B. Rimland, "High Dosage Levels of Certain Vitamins in the Treatment of Children with Severe Mental Disorders," in *Orthomolecular Psychiatry: Treatment of Schizophrenia*, ed. D. Hawkins and L. Pauling (San Francisco: W. H. Freeman and Company, 1973).

130   Child with autism dies from EDTA: K. Kane, "Death of 5-Year-Old Boy Linked to Controversial Chelation Therapy," *Pittsburgh Post-Gazette*, January 6, 2006.

131   Child with cerebral palsy dies from hyperbaric oxygen therapy: "Child Hurt in Chamber Explosion Dies in Hospital," CBS News, June 11, 2009.

131   Autism causes and treatments: J. McCarthy and J. Kartzinel, *Healing and Preventing Autism: A Complete Guide*. New York (Plume, 2010); K. Siri and T. Lyons, *Cutting-Edge Therapies for Autism: 2010-2011* (New York: Skyhorse Publishing, 2010); B. Jepson, *Changing the Course of Autism: A Scientific Approach for Parents and Physicians* (Boulder, Colo.: Sentient Publishing, 2007).

132   Bleach enemas: Orac, "The Lowest of the Low: Trying to Bleach Autism Away," *ScienceBlogs*, http://scienceblogs.com/insolence/2012/5/25/selling-bleach-as-a-cure-for-autism.

132   McCarthy blames MMR vaccine: McCarthy and Kartzinel, *Healing*.

132   McCarthy on vaccinating next child: Ibid., 278.

133   Oprah impressed with McCarthy: *The Oprah Winfrey Show*, September 18, 2007.

133   Autism research: Offit, *Autism's False Prophets: Bad Science, Risky Medicine, and the Search for a Cure* (New York: Columbia University Press, 2008).

134   Biomedical treatments don't work: J. H. Elder, M. Shanker, J. Shuster, et al., "The Gluten-Free, Casein-Free Diet in Autism: Results of a Preliminary Double Blind Clinical Trial," *Journal of Autism and Developmental Disorders* 36 (2006): 413–20; B. Jepson, D. Granpeesheh, J. Tarbox, et al. "Controlled Evaluation of the Effects of Hyperbaric Oxygen Therapy on the Behavior of 16 Children with Autism Spectrum Disorders," *Journal of Autism and Developmental Disorders* 41 (2011): 575–88; S. E. Soden, "24-Hour Provoked Urine Excretion Test for Heavy Metals in Children with Autism

and Typically Developing Children: A Pilot Study," *Clinical Toxicology* 45 (2007): 476–81.

134   Secretin: A. D. Sandler, K. A. Sutton, J. DeWeese, et al., "Lack of Benefit of a Single Dose of Synthetic Human Secretin in the Treatment of Autism and Pervasive Developmental Disorder," *New England Journal of Medicine* 341 (1999): 1801–1806; F. R. Volkmar, "Lessons from Secretin," *New England Journal of Medicine* 341 (1999): 1842–45; P. Sturmey, "Secretin Is an Ineffective Treatment for Pervasive Developmental Disabilities: A Review of 15 Double-Blind Randomized Controlled Trials," *Research in Developmental Disabilities* 26 (2005): 87–97.

135   Alison Singer: Alison Singer, interviewed by the author, September 26, 2011.

137   *American President* quote: *The American President*, directed by Rob Reiner (1995; Beverly Hills, Calif.: Castle Rock Entertainment, 1999).

138   Harm from biomedical treatments for autism: C. Plafki, P. Peters, M. Almeling, et al., "Complications and Side Effects of Hyperbaric Oxygen Therapy," *Aviation and Space Environmental Medicine* 71 (2000): 119–24; Centers for Disease Control and Prevention, "Deaths Associated with Hypocalcemia from Chelation Therapy—Texas, Pennsylvania, and Oregon, 2003–2005," *Morbidity and Mortality Weekly Report* 55 (2006): 204–207; National Institutes of Health, "Thin Bones Seen in Boys with Autism and Autism Spectrum Disorder," press release, January 29, 2008, www.nih.gov/news/health/jan2008/nichd-29.htm.

138   Harm caused by vaccine-preventable diseases: P. Offit, *Deadly Choices: How the Anti-Vaccine Movement Threatens Us All* (New York: Basic Books, 2011).

139   Alternative healers disdain vaccines: L. Downey, P. T. Tyree, C. E. Huebner, et al., "Pediatric Vaccination and Vaccine-Preventable Disease Acquisition: Associations with Care by Complementary and Alternative Medicine Providers," *Maternal and Child Health Journal* 14 (2010): 922–30.

139   Outbreaks of vaccine-preventable diseases: Offit, *Deadly Choices*.

Chapter 7: Chronic Lyme Disease

140   Gostin paper: J. D. Kraemer and L. O. Gostin, "Science, Politics, and Values: The Politicization of Professional Practice Guidelines," *Journal of the American Medical Association* 301 (2009): 665–67.

141   Dan Burton: B. Wilson, "The Rise and Fall of Laetrile," www.quackwatch.org/01QuackeryRelatedTopics/Cancer/laetrile.html; S. Brownlee,

"Swallowing Ephedra," archive.salon.com/health/feature/2000/06/07/ephedra; E. Walsh, "Burton: A 'Pit Bull' in the Chair," *Washington Post*, March 19, 1997; F. Pellegrini, "Fool on the Hill," *Time.com*, www.time.com/time/daily/special/look/burton.

141   Wakefield and MMR: Offit, *Autism's False Prophets*.

142   U. S. parents scared of MMR: M. J. Smith, S. S. Ellenberg, L. M. Bell, et al., "Media Coverage of the Measles-Mumps-Rubella Vaccine and Autism Controversy and Its Relationship to MMR Immunization Rates in the United States," *Pediatrics* 121 (2008): e836–e843.

142   Measles outbreaks in 2008: Offit, *Deadly Choices*, xv–xvi.

142   Measles outbreak in Europe: Centers for Disease Control and Prevention, "Measles Outbreaks," www.cdc.gov/measles/outbreaks.html.

142   Changing the value of *pi*: C. Seife, *Proofiness: The Dark Arts of Mathematical Deception* (New York: The Penguin Group, 2010).

143   Polly Murray and Old Lyme outbreak: P. Weintraub, *Cause Unknown: Inside the Lyme Epidemic* (New York: St. Martin's Griffin, 2008), 43–45.

143   Allen Steere and Lyme arthritis: A. C. Steere, S. E. Malawista, D. R. Snydman, et al., "Lyme Arthritis: An Epidemic of Oligoarticular Arthritis in Children and Adults in Three Connecticut Communities," *Arthritis and Rheumatism* 20 (1977): 7–17.

144   Willy Burgdorfer and *Borrelia burgdorferi*: W. Burgdorfer, A. G. Barbour, S. F. Hayes, et al., "Lyme Disease—A Tick-Borne Spirochetosis?" *Science* 216 (1982): 1317–19.

144   Lyme disease symptoms and treatment: G. P. Wormser, R. J. Dattwyler, E. D. Shapiro, et al., "The Clinical Assessment, Treatment, and Prevention of Lyme Disease, Human Granulocytic Anaplasmosis, and Babesiosis: Clinical Practice Guidelines by the Infectious Diseases Society of America," *Clinical Infectious Diseases* 43 (2006): 1089–1134.

145   Lyme bacteria cause other diseases: P. G. Auwaerter, J. S. Bakken, R. J. Dattwyler, et al., "Antiscience and Ethical Concerns Associated with Advocacy of Lyme Disease," *Lancet Infectious Diseases* 11 (2011): 713–19.

146   Chronic Lyme symptoms: C. Bean and L. Fein, *Beating Lyme: Understanding and Treating This Complex and Often Misdiagnosed Disease* (New York: AMACOM, 2008), 263–66.

146   Director's quote regarding *Under Our Skin*: *Under Our Skin*, Discussion Guide, Open Eye Productions, 2008.

147   *Under Our Skin*: Ibid.

147   Alternative medicine treatments for Chronic Lyme disease: Bean, *Beating Lyme*; S. Buhner, *Healing Lyme: Natural Healing and Prevention of Lyme Borreliosis and Its Coinfections* (Silver City, N.M.: Raven Press, 2005); N. McFadzean, *The Lyme Diet: Nutritional Strategies for Healing Lyme Disease* (San Diego: Legacy Line Publishing, 2010); G. Piazza and L. Piazza, *Recipes for Repair: A Lyme Disease Cookbook* (Sunapee, N.H.: Peconic Publishing, 2010); B. Rosner, *Lyme Disease and Rife Machines* (South Lake Tahoe, Calif.: BioMed Publishing Group, 2005); B. Rosner, *The Top 10 Lyme Disease Treatments: Defeat Lyme Disease with the Best of Conventional and Alternative Medicine* (South Lake Tahoe, Calif.: BioMed Publishing Group, 2007); K. Singleton, *The Lyme Disease Solution* (Charleston, S.C.: BookSurge Publishing, 2008); W. Storl, *Healing Lyme Disease Naturally: History, Analysis, and Treatments* (Berkeley, Calif.: North Atlantic Books, 2010); C. Strasheim, *Insights into Lyme Disease Treatment: 13 Lyme Literate Health Care Practitioners Share Their Healing Strategies* (South Lake Tahoe, Calif.: BioMed Publishing Group, 2009).

148   Wolf Storl and teasel: Storl, *Healing Lyme*.

149   Rife machine for Chronic Lyme: Rosner, *Top 10 Lyme Treatments*.

149   Rife machine testimonials: Rosner, *Lyme Disease and Rife Machines*: 165–85.

150   Killing Lyme bacteria: G. P. Wormser, R. J. Dattwyler, E. D. Shapiro, et al., "The Clinical Assessment, Treatment, and Prevention of Lyme Disease, Human Granulocytic Anaplasmosis, and Babesiosis: Clinical Practice Guidelines by the Infectious Diseases Society of America," *Clinical Infectious Diseases* 43 (2006): 1089–1134; Auwaerter, Bakken, Dattwyler, et al., "Antiscience and Ethical Concerns."

150   Chronic symptoms in Americans: Auwaerter, Bakken, Dattwyler, et al., "Antiscience and Ethical Concerns."

151   Lyme Literate doctors ignore treatable disorders: M. C. Reid, R. T. Schoen, J. Evans, et al., "The Consequences of Overdiagnosis and Overtreatment of Lyme Disease: An Observational Study," *Annals of Internal Medicine* 128 (1998): 354–62.

151   Evidence against the existence of Chronic Lyme disease: J. Radolf, "Posttreatment Chronic Lyme Disease—What It Is Not," *Journal of Infectious Diseases* 192 (2005): 948–49; G. P. Wormser, R. J. Dattwyler, E. D. Shapiro, et al., "The Clinical Assessment, Treatment, and

Prevention of Lyme Disease, Human Granulocytic Anaplasmosis, and Babesiosis: Clinical Practice Guidelines by the Infectious Diseases Society of America," *Clinical Infectious Diseases* 43 (2006): 1089–1134; P. Auwaerter, "Point: Antibiotic Therapy Is Not the Answer for Patients with Persisting Symptoms Attributable to Lyme Disease," *Clinical Infectious Diseases* 45 (2007): 143–48; H. M. Feder, B. J. B. Johnson, S. O'Connell, et al., "A Critical Appraisal of 'Chronic Lyme Disease,' " *New England Journal of Medicine* 357 (2007): 1422–30; A. Marques, "Chronic Lyme Disease: A Review," *Infectious Disease Clinics of North America* 22 (2008): 341–60; Auwaerter, Bakken, Dattwyler, et al., "Antiscience and Ethical Concerns."

151 *New England Journal of Medicine* paper: H. M. Feder, B. J. B. Johnson, S. O'Connell, et al., "A Critical Appraisal of 'Chronic Lyme Disease,' " *New England Journal of Medicine* 357 (2007): 1422–30.

151 Woman dies from blood clot in heart: R. Patel, K. L. Grogg, W. D. Edwards, et al., "Death from Inappropriate Therapy for Lyme Disease," *Clinical Infectious Diseases* 31 (2000): 1107–1109.

152 Outbreak of gallstones in New Jersey: Centers for Disease Control and Prevention, "Ceftriaxone-Associated Biliary Complications of Treatment of Suspected Disseminated Lyme Disease—New Jersey, 1990–1992," *Morbidity and Mortality Weekly Report* 42 (1993): 39–42; P. J. Ettestad, G. L. Campbell, S. F. Welbel, et al., "Biliary Complications in the Treatment of Unsubstantiated Lyme Disease," *Journal of Infectious Diseases* 171 (1995): 356–61.

152 Other complications from prolonged antibiotic therapy: M. C. Reid, R. T. Schoen, J. Evans, et al., "The Consequences of Overdiagnosis and Overtreatment of Lyme Disease: An Observational Study," *Annals of Internal Medicine* 128 (1998): 354–62.

152 Malariatherapy for Chronic Lyme disease: Centers for Disease Control and Prevention, "Epidemiologic Notes and Reports: Imported Malaria Associated with Malariotherapy of Lyme Disease—New Jersey," *Morbidity and Mortality Weekly Report* 39 (1990): 873–75.

152 Complications from bismuth therapy: U. S. Food and Drug Administration, "FDA Warns Consumers and Health Care Providers Not to Use Bismacine, Also Known as Chromacine," press release, July 21, 2006.

152 Misbehavior of Lyme Literate doctors: Auwaerter, Bakken, Dattwyler, et al., "Antiscience and Ethical Concerns."

153 Statement by IDSA against alternative Lyme therapies: G. P. Wormser, R. J. Dattwyler, E. D. Shapiro, et al., "The Clinical Assessment, Treatment, and Prevention of Lyme Disease, Human Granulocytic Anaplasmosis, and Babesiosis: Clinical Practice Guidelines by the Infectious Diseases Society of America," *Clinical Infectious Diseases* 43 (2006): 1089–1134.

153 ILADS guidelines: D. Cameron, A. Gaito, N. Harris, et al., "ILADS Working Group: Evidence-Based Guidelines for the Management of Lyme Disease," *Expert Reviews of Anti-Infective Therapy* 2 (2004): S1–S13.

153 Lyme activist groups recruit politicians: B. Patoine, "Guideline-Making Gets Tougher: Action by State Attorney General Over Lyme Disease Guidelines Stirs Debate," *Annals of Neurology* 65 (2009): A10–A15.

154 Blumenthal's support of Lyme activists: P. G. Auwaerter, J. S. Bakken and R. J. Dattwyler, "Scientific Evidence and Best Patient Care Practices Should Guide the Ethics of Lyme Disease Activism," *Journal of Medical Ethics* 37 (2011): 1–6.

154 Jones and stretch limo: D. Whelan, "Lyme Inc.," *Forbes*, March 12, 2007, www.forbes.com/forbes/2007/0312/096_print.html.

154 Blumenthal sues IDSA: Office of the Attorney General, "Attorney General's Investigation Reveals Flawed Lyme Disease Guideline Process, IDSA Agrees to Reassess Guidelines, Install Independent Arbiter," press release, May 1, 2008, www.policymed.com/2010/05/richard-blumenthals-lyme-deception.html.

155 AAN attorneys argue case: B. Patoine, "Guideline-Making Gets Tougher."

155 Rulings by FTC and DOJ: *Schachar v. American Academy of Ophthalmology Inc.* 870 F2d 397 (7th Circuit, 1989).

155 Blumenthal claims conflict of interest: Office of the Attorney General, "Attorney General's Investigation Reveals Flawed Lyme Disease Guideline Process."

155 IDSA panel doesn't benefit from guidelines: "Richard Blumenthal's Lyme Deception," *Policy and Medicine*, May 18, 2010, www.mdjunction.com/forums/lyme-disease-support-forums/lyme-disease-activism/1691814-richard-blumenthals-lyme-deception-51810.

155 ILADS and Lyme diagnostic test maker: J. D. Kraemer and L. O. Gostin, "Science, Politics, and Values: The Politicization of Professional Practice Guidelines," *Journal of the American Medical Association* 301 (2009): 665–67.

155 Conflict with intravenous infusion companies: Auwaerter, Bakken, and Dattwyler, "Scientific Evidence and Best Patient Care Practices."

156   Chronic Lyme doctor and insurance company: Whelan, "Lyme Inc."

156   Chronic Lyme doctor and waterfall: Ibid.

156   Chronic Lyme doctor in California: P. Auwaerter, Bakken, and Dattwyler, "Scientific Evidence and Best Patient Care Practices"; Whelan, "Lyme Inc."

156   IDSA litigation costs: J. D. Kraemer and L. O. Gostin, "Science, Politics, and Values: The Politicization of Professional Practice Guidelines," *Journal of the American Medical Association* 301 (2009): 665–67.

156   Anne Gershon quote: Anne Gershon, interviewed by the author, October 7, 2011.

156   Howard Brody: "Richard Blumenthal's Lyme Deception."

157   IDSA final report: P. M. Lantos, W. A. Charini, G. Medoff, et al., "Final Report of the Lyme Disease Review Panel of the Infectious Diseases Society of America," *Clinical Infectious Diseases* 51 (2010): 1–5.

157   Blumenthal and 75 percent requirement: "Richard Blumenthal's Lyme Deception."

157   Anger at $10,000 cutoff: Ibid.; B. Patoine, "Guideline-Making Gets Tougher."

157   Carol Baker quotes: Carol Baker, interviewed by the author, September 26, 2011.

158   Blumenthal response to IDSA's final report: Attorney General Statement on IDSA Guidelines Review Panel Report, www.ct.gov/ag/cwp/view.asp?A=2341&Q=459296.

158   401(k) account story: P. Callahan and T. Tsouderos, "Chronic Lyme Disease: A Dubious Diagnosis," *Chicago Tribune*, December 8, 2010.

158   Gostin and Kraemer regarding scientific process: J. D. Kraemer and L. O. Gostin, "Science, Politics, and Values: The Politicization of Professional Practice Guidelines," *Journal of the American Medical Association* 301 (2009): 665–67.

CHAPTER 8: CURING CANCER

163   Steve Jobs: P. Elkind, "The Trouble with Steve Jobs," *Fortune*, March 5, 2008, http://money.cnn.com/2008/03/02/news/companies/elkind_jobs.fortune/index.htm?postversion=2008030510; S. Begley, "Jobs's Unorthodox Treatment," *The Daily Beast*, October 5, 2011, http://www.thedailybeast.com/articles/2011/10/05/steve-jobs-dies-his-unorthodox-treatment-for-neuro endocrine-cancer.html; S. Lohr, "Jobs Tried Exotic Treatments to Combat

Cancer, Book Says," *New York Times*, October 20, 2011; W. Isaacson, *Steve Jobs* (New York: Simon & Schuster, 2011): 454.

164 Early American cancer cures: Anonymous, *Nostrums and Quackery: Articles on the Nostrum Evil and Quackery Reprinted, with Additions and Modifications, from the Journal of the American Medical Association* (Chicago: American Medical Association Press, 1912), 25–75; Young, *American Health Quackery*, 234; Barrett, *Consumer Health*, 370, 375.

164 Abrams and charlatan: Young, *American Health Quackery*, 189.

164 Albert Abrams story: Young, *American Health Quackery*, 189; C. Jameson, *The Natural History of Quackery* (Springfield, Ill.: Charles C. Thomas, 1961), 210–12; P. Modde, *Chiropractic Malpractice* (Columbia, Md.: Henrow Press, 1985), 103; R. A. Lee, *The Bizarre Careers of John R. Brinkley* (Lexington: The University Press of Kentucky, 2002), xii–xiii; McCoy, *Quack!*, 72–83; Young, *Medical Messiahs*, 137–42; M. Fishbein, *Fads and Quackery in Healing: An Analysis of the Foibles of the Healing Cults, with Essays on Various Other Peculiar Notions in the Health Field* (New York: Blue Ribbon Books, 1932), 140–55; M. Fishbein, *The Medical Follies* (New York: Boni and Liveright, 1925), 99–118.

165 Robert Millikan quote: Ibid.

166 *Scientific American* editor's quote: Young, *Medical Messiahs*, 140.

166 Glyoxylide: Young, *American Health Quackery*, 235; Barrett, *Health Schemes*, 5; Barrett and Jarvis, *Health Robbers*, 27, 93.

166 Harry Hoxsey: Young, *American Health Quackery*, 235; Johnston, *Politics of Healing*, 235–36; Barrett, *Consumer Health*, 370; Young, *Medical Messiahs*, 360–89; Juhne, *Quacks and Crusaders*, 64–91.

167 " 'Dem that gets took" quote: Young, *Medical Messiahs*, 363.

168 Career of Andrew Ivy: M. I. Grossman, "Andrew Conway Ivy (1893–1978)," *Physiologist* 21 (1978): 11–12; D. B. Dill, "A. C. Ivy—Reminiscences," *Physiologist* 22 (1979): 21–22; E. Shuster, "Fifty Years Later: The Significance of the Nuremberg Code," *New England Journal of Medicine* 337 (1997): 1436–40.

169 Moreno regarding Ivy: J. D. Moreno, *Undue Risk: Secret State Experiments on Humans* (New York: Routledge, 2001), 66.

169 Krebiozen: Barrett, *Consumer Health*, 370; J. F. Holland, "The Krebiozen Story," *Quackwatch*, www.quackwatch.com/01QuackeryRelatedTopics/Cancer/krebiozen.html; S. Chertow, "Krebiozen," The Chicago Literary Club, www.chilit.org/Papers%20by%20author/Chertow%20--%20

Krebiozen.HTM; W. F. Janssen, "Cancer Quackery: Past and Present," Cancer Treatment Watch, www.cancertreatmentwatch.org/q/janssen .shtml; P. S. Ward, "Who Will Bell the Cat? Andrew C. Ivy and Krebiozen," *Bulletin of the History of Medicine* 58 (1984): 28–52; A. C. Ivy, *Krebiozen: An Agent for the Treatment of Malignant Tumors* (Chicago: Champlin-Shealy Company, 1951).

170   Gerson diet: Barrett and Jarvis, *Health Robbers*, 87–88; Barrett, *Consumer Health*, 372; K. Butler, *A Consumer's Guide to "Alternative Medicine": A Close Look at Homeopathy, Acupuncture, Faith-Healing, and Other Unconventional Treatments* (Amherst, N.Y.: Prometheus Books, 1992), 43; S. Barrett, "Questionable Cancer Therapies," *Quackwatch,* www.quackwatch. com/01QuackeryRelatedTopics/cancer.html.

171   Shark cartilage: Barrett, *Consumer Health,* 371–72; Barrett, *Vitamin Pushers*, 374–76; Hurley, *Natural Causes*, 201–204, 223; Shapiro, *Suckers*, 176–78; Wanjek, *Bad Medicine*, 103–107; Singh and Ernst, *Trick or Treatment*, 263–64; M. J. Coppes, R. A. Anderson, R. M. Egeler, and J. E. A. Wolff, "Alternative Therapies for the Treatment of Childhood Cancer," *New England Journal of Medicine* 339 (1998): 846–47.

CHAPTER 9: SICK CHILDREN, DESPERATE PARENTS

174   Billie Bainbridge: R. Smith, "Mum and Four-Year-Old Daughter Both Battling Cancer," *Daily Mirror*, August 8, 2011; " 'A Lot of People Ask Me How I Cope, But You Just Kind of Deal With It—That's All You Can Do,' " *Exeter Express and Echo*, August 13, 2011; C. Axford and L. French, "Exeter Family Hit Twice by Cancer 'Fights On' for Child," BBC News, August 19, 2011; A. Radnedge, "Mother and Daughter, 4, in Fight to Battle Cancer," *Metro*, August 21, 2011; "Billie Bainbridge Set to Fly to Take Part in Medical Trial," *Exeter Express and Echo*, September 15, 2011; "Positive Start for Billie Bainbridge's America Treatment," *Exeter Express and Echo*, September 29, 2011; "Brave Billie Starting Treatment in States," *Exeter Express and Echo*, October 6, 2011; "Billie Fund Rises Toward Target," *Exeter Express and Echo*, October 13, 2011; "Chiefs Added to Growing Support for Billie's Fund," *Exeter Express and Echo*, October 20, 2011; "Raffle Prize Goes to Butterfly Fund," *Exeter Express and Echo*, November 3, 2011; L. Bainbridge, "The Worst Year of My Life: Cancer Has My Family in Its Grip," *The Observer*, November 19, 2011.

175 Burzynski's early research career: M. E. G. Smith, "The Burzynski Controversy in the United States and Canada: A Comparative Case Study in the Sociology of Alternative Medicine," *The Canadian Journal of Sociology* 17 (1992): 133–60.

176 Burzynski defines antineoplastons: S. R. Burzynski, "Antineoplastons: Biochemical Defense Against Cancer," *Physiological Chemistry and Physics* 8 (1976): 275–79.

177 Burzynski Research Institute: M. E. G. Smith, "The Burzynski Controversy in the United States and Canada: A Comparative Case Study in the Sociology of Alternative Medicine," *The Canadian Journal of Sociology* 17 (1992): 133–60.

177 Burzynski collects urine: G. Null, "The Suppression of Cancer Cures," *Penthouse*, October 1979; "War on Cancer: Politics or Profit," *20/20*, October 21, 1981.

177 Null article: Null, "The Suppression of Cancer Cures."

178 Geraldo Rivera report: "War on Cancer: Politics or Profit," *20/20*, October 21, 1981.

178 Patients cured by antineoplastons: T. Elias, *The Burzynski Breakthrough* (Nevada City, Calif.: Lexikos, 2001); G. Phillips, "Interview with Dr. Burzynski," December 5, 2003, www.cancerinform.org/aburzinterview.html.

179 Harry Smith interviews Burzynski's patients: D. Fehling, "Controversial Cancer Doc: Urine Treatment Works," KENS 5-TV, www.kens5.com/archive/66499497.html.

180 Burzynski movie: *Burzynski: The Movie*, directed by Eric Merola (Merola Films, 2010).

180 Burzynski Research Institute: Ibid.

181 Blackstein and Bergasel and "hoodwinked": Barrett, *Consumer Health*, 373; M. E. Blackstein and D. E. Bergsagel. *Report to the Ontario Ministry of Health on the Treatment of Cancer Patients with Antineoplastons and the Burzynski Clinic in Houston, Texas.* Undated, circa 1983.

181 Blackstein, Bergasel, and additional patients: "Pharmacologic and Biological Treatments," in *Unconventional Cancer Treatments*, *Quackwatch*, www.quackwatch.org/01QuackeryRelatedTopics/OTA/ota05.html.

182 1985 review: Ibid.

182 Sally Jessy Raphael and *Inside Edition*: Barrett, *Consumer Health*, 373.

182 Office of Technology Assessment: American Cancer Society, "Anti-Neoplaston Therapy," www.cancer.org/Treatment/TreatmentsandSide

Effects/ComplementaryandAlternativeMedicine/Pharmacologicaland BiologicalTreatment/antineoplaston-therapy.

183   Jacobs regarding clowns: P. Goldberg, "The Antineoplaston Anomaly: How a Drug was Used for Decades in Thousands of Patients with No Safety, Efficacy Data," *The Cancer Letter*, September 25, 1998.

183   National Cancer Institute trial of antineoplastons: J. Buckner, M. Malkin, E. Reed, et al., "Phase II Study of Antineoplastons A10 (NSC 648539) and AS2-1 (NSC 620261) in Patients with Recurrent Glioma," *Mayo Clinic Proceedings* 74 (1999): 137–45.

183   Burzynski angry at NCI study: G. Phillips, "Interview with Dr. Burzynski," December 5, 2003, www.cancerinform.org/aburzinterview.html.

184   Saul Green reviews antineoplastons: S. Green, "Antineoplastons: An Unproved Cancer Therapy," *Journal of the American Medical Association* 267 (1992): 2924–28.

184   Review of Burzynski by cancer specialists: Goldberg, "The Antineoplaston Anomaly."

185   Congressional hearing: *Burzynski: The Movie*.

185   Cancer activists angry at Burzynski: Goldberg, "The Antineoplaston Anomaly."

186   Sigma Tau Pharmaceuticals drops antineoplaston research: Green, "Antineoplastons: An Unproved Cancer Therapy."

186   Burzynski claims conspiracy among oncologists: Phillips, "Interview with Dr. Burzynski."

186   Goldberg and Friedman regarding cancer drug testing: Goldberg, "The Antineoplaston Anomaly."

187   PLX4032: A. Harmon, "A Roller Coaster Chase for Cure," *New York Times*, February 21, 2010; A. Harmon, "New Drugs Stir Debate on Rules of Clinical Trials," *New York Times*, September 18, 2010; A. Harmon, "Drug to Fight Melanoma Prolonged Life in Trial," *New York Times*, January 19, 2011; R. Schwartz, "*The Emperor of All Maladies*—The Beginning of the Beginning," *New England Journal of Medicine* 365 (2011): 2353–55.

188   Adamson regarding Burzynski: Peter Adamson, interviewed by the author, September 19, 2011.

188   Burzynski as caring man: C. Malisow, "Cancer Doctor Stanislaw Burzynski Sees Himself as a Crusading Researcher, Not a Quack," *Houston Post*, January 1, 2009.

189   Adamson regarding false hope: Peter Adamson, interviewed by the author, September 19, 2011.

189 Maris regarding false hope: John Maris, interviewed by the author, September 28, 2011.

190 Jeanine Graf regarding Burzynski's therapies: Jeanine Graf, interviewed by the author, March 1, 2012.

191 Science blogger regarding false hope: Andy Lewis, "The False Hope of the Burzynski Clinic," November 21, 2011, www.quackometer.net/blog/2011/11/the-false-hope-of-the-burzynski-clinic.html.

191 Billie Bainbridge dies: "Billie Bainbridge Dies After Battle with Brain Stem Cancer," BBC News, June 5, 2012, www.bbc.co.uk/news/uk-england-devon-18331017.

192 Burzynski's anti-aging medicines: www.aminocare.com.

192 Somers's book: S. Somers, *Knockout: Interviews with Doctors Who Are Curing Cancer and How to Prevent Getting It in the First Place* (New York: Crown Publishing Group, 2009).

192 Mukherjee's book: S. Mukherjee, *The Emperor of All Maladies: A Biography of Cancer* (New York: Scribner, 2010).

CHAPTER 10: MAGIC POTIONS IN THE
TWENTY-FIRST CENTURY

198 "One for the Angels": *The Twilight Zone,* episode 2, "One for the Angels" (Los Angeles: Cayuga Productions, October 9, 1959).

200 Origins of *quack*: Barrett and London, *Consumer Health*, 36.

201 Patent medicines: Anonymous, *Nostrums and Quackery*; Cramp, *Nostrums and Quackery*.

201 Kickapoo Joy Juice: McNamara, *Step Right Up*, xiv.

201 Celebrity supporters: McCoy, *Quack*, 95, 201.

202 Buttar's background: Buttar, *9 Steps*, 329–31.

203 Buttar regarding man-made harms: Ibid., 81.

203 Risks of environmental toxins are overstated: G. Kabat, *Hyping Health Risks: Environmental Hazards in Daily Life and the Science of Epidemiology* (New York: Columbia University Press, 2011).

203 Chelation studies fail to show effect: E. Ernst, "Chelation Therapy for Peripheral Arterial Occlusive Disease: A Systematic Review," *Circulation* 96 (1997): 1031–33.

204 Garcia testimony: *In the Matter of Rashid Ali Buttar, D. O., Before the North Carolina Medical Board*, April 24, 2008 (testimony by Jane Garcia).

204   Buttar on son's birth: Rashid A. Buttar, D. O., *Know Your Options: Autism: The Misdiagnosis of Our Future Generations*, DVD (Coral Gables, Fla.: Dolphin Entertainment, 2006).

204   Buttar regarding mission from God: Buttar, *9 Steps*, 264.

205   Buttar on struggle with God: Ibid., 264–65.

205   Buttar on son's recovery: Ibid.

205   Jennings's symptoms: J. Avila and D. Cohen, "Medical Mystery or Hoax: Did Cheerleader Fake a Muscle Disorder?" ABC News, July 23, 2010; "Desiree Jennings," *RationalWiki*, http://rationalwiki.org/wiki/Desiree_Jennings.

206   Jennings and Coldplay: Ibid.

206   Jennings and British accent: "Flu Shot Woman," *Inside Edition*, http://www.insideedition.com/headlines/159-flu-shot-woman.

206   Jennings's diagnostic tests: "The Flu, a Shot to the System," www.loudun-times.com/index.php/archive/article/Column_The_flu_a_shot_to_the_system.

206   Jennings and dystonia: Ibid.

206   *Inside Edition* story: "Woman Says Flu Shot Triggered Rare Neurological Disorder," www.wusa9.com/news/local/story.aspx?storyid=92345&catid=158.

206   Jenny McCarthy directs Jennings to Buttar: "Desiree Jennings Update: Road to Recovery," *Planet Thrive*, http://planetthrive.com/2009/11/desiree-jennings-update-road-to-recovery.

207   Buttar diagnoses Jennings with mercury toxicity: C. Coffey, "Woman Disabled by Flu Shot Reaction," www.myfoxdc.com/dpp/health/101309_woman_disabled_by_flu_shot_reaction_dystonia.

207   Jennings on *20/20*: Avila and Cohen, "Medical Mystery or Hoax."

207   Hopkins neurologist diagnoses psychogenic disorder: S. Novella, "Desiree Jennings: The Plot Thickens," *Neurologica* blog, http://theness.com/neurologicablog/?p=1558; Orac, "Has Desiree Jennings VAERS Report Been Found?" *ScienceBlogs*, http://scienceblogs.com/insolence/2009/11/05/has-desiree-jennings-vaers-report-been-f.

207   Novella regarding Jennings: S. Novella, "Desiree Jennings on *20/20*," *Neurologica* blog, http://theness.com/neurologicablog/?p=2150.

207   University of Maryland uses Jennings's tape to train residents: "Desiree Jennings," *RationalWiki*; http://www.foxnews.com/search-results/m/26952743/flu-shot-fears.htm.

207   *Inside Edition* follow-up story: "Flu Shot Woman," *Inside Edition*.

208   Novella regarding Jennings: www.theness.com/neurologicablog/?p=1195.

208   Jennings and China: Avila and Cohen, "Medical Mystery or Hoax."

208   Problems with provoked metals excretion testing: S. Barrett, "How the 'Urine Toxic Metals' Test Is Used to Defraud Patients," *QuackWatch*, www .quackwatch.org/01QuackeryRelatedTopics/Tests/urine_toxic.html (italics mine).

208   Buttar testifies at congressional hearing: *Autism Spectrum Disorders: An Update of Federal Government Initiatives and Revolutionary New Treatments of Neurodevelopmental Diseases, Before the Subcommittee on Human Rights & Wellness Government Reform Committee*, May 6, 2004 (testimony by Rashid A. Buttar).

209   Heavy-metals test on autistic and normal children: S. E. Soden et al., "24-Hour Provoked Urine Excretion Test for Heavy Metals in Children with Autism and Typically Developing Children: A Pilot Study," *Clinical Toxicology* 45 (2007): 476–81.

209   Novella regarding Jennings: S. Novella, "Well That Didn't Take Long—Another Dystonia Case Follow Up," *Neurologica* blog, www.theness.com/neurologicablog/?p=1195.

209   Buttar regarding testing of TD-DMPS: Estherar, "The Allure of Biomedical Treatments [*sic*] for Autism," Mainstream Parenting Resources, http://mainstreamparenting.wordpress.com/2009/05/21/the-allure-of-biomedical-treatments-for-autism.

211   Buttar regarding the value of TD-DMPS: Ibid.

211   Buttar warns against using other transdermal chelation therapies: Rashid A. Buttar, "Buttar Autism Treatment Protocol: Advanced Concepts in Medicine/Center for Advanced Medicine," Center for Advanced Medicine (August 2004).

211   Buttar regarding Trans-D: Buttar, *9 Steps*, 254.

211   Older cure-alls: Anonymous, *Nostrums*, 436–53.

212   Sales of Trans-D: Buttar, *9 Steps*, 238.

212   Buttar regarding pharmaceutical companies: Rashid A. Buttar, D.O., *Know Your Options: Sudden Cardiac Death, #1 Symptom of Heart Disease*, DVD (Coral Gables, Fla.: Dolphin Entertainment, 2007).

212   Specter on Big Placebo: "Michael Specter: The Danger of Science Denial," YouTube, www.youtube.com/watch?v=7OMLSs8t1ng.

212   Buttar's oath: Buttar, *9 Steps*, xvii.

212 Buttar warns against medical doctors: Ibid., 10.

213 Buttar insists on unquestioning compliance: Rashid A. Buttar, D.O., *Know Your Options: Cancer, The Untold Truth,* DVD (Coral Gables, Fla.: Dolphin Entertainment, 2007).

213 Buttar claims that doctors don't follow their own advice: Ibid.

213 Buttar and CDC conspiracy: Rashid A. Buttar, D. O., *Know Your Options: Heavy Metal Toxicity: The Hidden Killer,* DVD (Coral Gables, Fla.: Dolphin Entertainment, 2007).

214 CDC warning about chelation therapy: "Deaths Associated with Hypocalcemia from Chelation Therapy—Texas, Pennsylvania, and Oregon, 2003–2005," *Morbidity and Mortality Weekly Report* 55 (2006): 204–7 (italics mine).

215 Buttar on how his son became autistic: Buttar, *9 Steps,* 263.

215 Buttar regarding thimerosal: Rashid A. Buttar, D. O., *Know Your Options: Heavy Metal Toxicity.*

215 Buttar on not vaccinating his child: Ibid.

215 Studies showing that thimerosal in vaccines didn't cause autism or mercury toxicity: Institute of Medicine, *Immunization Safety Review: Thimerosal-Containing Vaccines and Neurodevelopmental Disorders* (Washington, D. C.: National Academies Press, 2001); K. M. Madsen, M. B. Lauritsen, C. B. Pedersen, et al., "Thimerosal and the Occurrence of Autism: Negative Ecological Evidence from Danish Population-Based Data," *Pediatrics* 112 (2003): 604–6; A. Hviid, M. Stellfeld, J. Wohlfahrt and M. Melbye, "Association Between Thimerosal-Containing Vaccine and Autism," *Journal of the American Medical Association* 290 (2003): 1763–66; J. Heron and J. Golding, "Thimerosal Exposure in Infants and Developmental Disorders: A Prospective Cohort Study in the United Kingdom Does Not Support a Causal Association," *Pediatrics* 114 (2004): 577–83; N. Andrews, E. Miller, A. Grant, et al., "Thimerosal Exposure in Infants and Developmental Disorders: A Retrospective Cohort Study in the United Kingdom Does Not Support a Causal Association," *Pediatrics* 114 (2004): 584–91; P. Stehr-Green, P. Tull, M. Stellfeld, et al., "Autism and Thimerosal-Containing Vaccines: Lack of Consistent Evidence for an Association," *American Journal of Preventive Medicine* 25 (2005): 101–6; E. Fombonne, R. Zakarian, A. Bennett, et al., "Pervasive Developmental Disorders in Montreal, Quebec, Canada: Prevalence and Links with Immunization," *Pediatrics* 118 (2006): 139–50; W. W. Thompson, C. Price, B. Goodson, et al., "Early Thimerosal Exposure and Neuropsychological Outcomes at 7 to 10 Years," *New England Journal of Medicine* 357 (2007): 1281–92; R. Schechter

and J. Grether, "Continuing Increases in Autism Reported to California's Development Services System," *Archives of General Psychiatry* 65 (2008): 19–24; E. Fombonne, "Thimerosal Disappears but Autism Remains," *Archives of General Psychiatry* 65 (2008): 15–16.

215   Smallpox epidemiology: S. Plotkin, W. Orenstein, and P. Offit, eds., *Vaccines*, 6th ed. (Philadelphia: Saunders, 2012), 719–22.

215   Polio epidemiology: Centers for Disease Control and Prevention: "Outbreaks Following Wild Poliovirus Importations—Europe, Africa, and Asia, January 2009–September 2010," *Morbidity and Mortality Weekly Report* 59 (2010): 1393–99.

216   Hepatitis B virus epidemiology: G. L. Armstrong, E. F. Mast, M. Wojczynski and H. S. Margolis, "Childhood Hepatitis B Virus Infections in the United States Before Hepatitis B Immunization," *Pediatrics* 108 (2001): 1123–28.

216   Buttar warns against swine flu vaccine: "Dr. Rashid A. Buttar Speaks on the N1H1 [*sic*] Swine Flu Vaccine," www.youtube.com/watch?v=qcclPdW-mzmk, October 5, 2009.

216   Swine flu epidemiology: "2009 H1N1 Flu: Situation Update," May 28, 2010, CDC.gov, http://cdc.gov/h1n1flu/update.htm.

216   Swine flu vaccine safety: "Safety of Influenza A (H1N1) 2009 Monovalent Vaccines—United States, October 1–November 24, 2009," *Morbidity and Mortality Weekly Report* 58 (2009): 1–6.

217   Buttar trial before North Carolina Board of Medical Examiners: *In the Matter of Rashid Ali Buttar, D. O., Before the North Carolina Medical Board*, April 24, 2008 (testimony by Rashid Buttar).

218   Buttar asked to change consent form: S. Barrett, "Rashid Buttar Reprimanded," *Casewatch*, www.casewatch.org/board/med/buttar/consent_2010.shtml.

Chapter 11: The Remarkably Powerful, Highly Underrated Placebo Response

224   Acupuncture studies: Singh and Ernst, *Trick or Treatment*, 77–88; P. White, F. L. Bishop, P. Prescott, et al., "Practice, Practitioner, or Placebo? A Multifactorial, Mixed-Methods Randomized Controlled Trial of Acupuncture," *Pain: Journal of the International Association for the Study of Pain*, December 12, 2011, doi: 10.1016/jpain 2011.11.007.

224 Novella background: D. Gorski, "The Trouble with Dr. Oz," *Science-Based Medicine* blog, www.sciencebasedmedicine.org/?p=12208.

224 Oz regarding Novella's opinions on acupuncture: "Controversial Medicine: Alternative Health," *The Dr. Oz Show,* www.doctoroz.com/videos/alternative-medicine-controversy-pt-1.

224 Novella regarding acupuncture: Ibid.

225 Nurse and morphine: D. A. Lessing, K. Madden, S. Marlan, eds., *Encyclopedia of Psychology and Religion* (New York: Springer Publishing Co., 2009).

225 Unicorn horns: A. Shapiro and E. Shapiro, *The Powerful Placebo: From Ancient Priest to Modern Physician* (Baltimore: Johns Hopkins University Press, 1997), 14.

225 Vitamin O: R. Park, *Voodoo Science: The Road from Foolishness to Fraud* (Oxford: Oxford University Press, 2000), 46–49.

227 Therapist's attitude and placebo effect: W. G. Thompson, *The Placebo Effect and Health: Combining Science and Compassionate Care* (Amherst, N.Y.: Prometheus Books, 2005), 182.

227 Kaptchuk regarding personality: M. Specter, "The Power of Nothing," *The New Yorker*, December 12, 2011.

227 Blau regarding pathologist: Park, *Voodoo Science*, 49.

228 John Diamond regarding alternative medicine: Diamond, *Snake Oil*, 34.

229 *Postcards from the Edge* quote: *Postcards from the Edge*, directed by Mike Nichols (1990; Culver City, Calif.: Sony Pictures Home Entertainment, 2001).

230 Levine study: J. D. Levine, N. C. Gordon, and H. L. Fields, "The Mechanism of Placebo Analgesia," *Lancet* 312 (1978): 654–57.

230 Endorphin findings reproduced: F. Benedetti, *Placebo Effects: Understanding the Mechanisms in Health and Disease* (Oxford: Oxford University Press, 2009), 63–98.

231 Acupuncture and endorphins: V. Clement-Jones, L. McLoughlin, S. Tomlin, et al., "Increased Beta-Endorphin But Not Met-Enkephalin Levels in Human Cerebrospinal Fluid After Acupuncture for Recurrent Pain," *Lancet* 2 (1980): 946–49.

232 Caplan regarding ethics of deception in medicine: Art Caplan, interviewed by the author, September 20, 2011.

232 Caplan regarding deceptions by mainstream clinicians: Ibid.

233 Racetrack study: R. E. Knox and J. A. Inkster, "Postdecision Dissonance at Post Time," *Journal of Personality and Social Psychology* 8 (1968): 319–23.

233 *JAMA* pain study: R. L. Waber, B. Shiv, Z. Carmon, and D. Ariely, "Commercial Features of Placebo and Therapeutic Efficacy," *Journal of the American Medical Association* 299 (2008): 1016–17.

233 Caplan on cost of placebo response: Art Caplan, interviewed by the author, September 20, 2011.

233 South Korean president: "Acupuncture Needle Removed from Lung of Former South Korean President," *The Telegraph*, May 3, 2011.

233 Dangers of acupuncture: E. Ernst, "Deaths After Acupuncture: A Systematic Review," *International Journal of Risk and Safety in Medicine* 22 (2010): 131–36; "Acupuncture Can Spread Serious Diseases: Experts," Reuters, March 18, 2010, www.reuters.com/article/2010/03/19/us-acupuncture-infection-idUSTRE62I00220100319; I. Sample, "Dozens Killed by Incorrectly Placed Acupuncture Needles," *The Guardian*, October 18, 2010.

234 Ader-Cohen study: R. Ader and N. Cohen, "Behaviorally Conditioned Immunosuppression," *Psychosomatic Medicine* 37 (1975): 333–40.

235 MacKenzie study: J. N. MacKenzie, "The Production of the So-Called 'Rose-Cold' by Means of an Artificial Rose," *American Journal of the Medical Sciences* 91 (1886): 45–57.

235 Imboden study: J. B. Imboden, A. Canter and L. E. Cluff, "Convalescence from Influenza: A Study of the Psychological and Clinical Determinants," *Archives of Internal Medicine* 108 (1961): 393–99.

236 Ader lupus study: K. Olness and R. Ader, "Conditioning as an Adjunct in the Pharmacotherapy of Lupus Erythematosis," *Journal of Developmental and Behavioral Pediatrics* 13 (1992): 124–25.

236 Sabbioni study: M. E. E. Sabbioni, D. H. Bovbjerg, S. Mathew, et al., "Classically Conditioned Changes in Plasma Cortisol Levels Induced by Dexamethasone in Healthy Men," *FASEB Journal* 11 (1997): 1291–96.

236 Enhancing immune response study: D. L. Longo, P. L. Duffy, W. C. Koop, et al., "Conditioned Immune Response to Interferon-Gamma in Humans," *Clinical Immunology* 90 (1999): 173–81.

236 Benefits of placebo for variety of diseases: Benedetti, *Placebo Effects*; S. S. Wang, "Why Placebos Work Wonders," *Wall Street Journal*, January 3, 2012.

237 Homeopathy on Oz show: "Alternative Pain Treatments, Part 1," *The Dr. Oz Show*, www.doctoroz.com/videos/Alternative-Pain-Treatments-Pt-1.

238 Studies of homeopathic remedies: E. Ernst, "Homeopathy: What Does the 'Best' Evidence Tell Us?" *Medical Journal of Australia* 192 (2010): 458–60.

239　Harm caused by over-the-counter cough-and-cold preparations: D. Childs, "Docs Support FDA Cough Medicine Warning," ABC News, August 16, 2007, http://abcnews.go.com/Health/Drugs/story?id=3488351&page=1.

239　Ebers Papyrus: Shapiro and Shapiro, *Powerful Placebo*, 4.

240　Norman Cousins quote: N. Cousins, *Anatomy of an Illness* (New York: W. W. Norton & Company, 1979), 76–78.

CHAPTER 12: WHEN ALTERNATIVE MEDICINE BECOMES QUACKERY

241　Childhood deaths and prayer: C. Fraser, *God's Perfect Child: Living and Dying in the Christian Science Church* (New York: Henry Holt and Company, 1999).

241　Weil regarding HIV treatments: A. Weil, *Spontaneous Healing: How to Discover and Enhance Your Body's Natural Ability to Maintain and Heal Itself* (New York: Balantine Books, 1995), 4.

242　Benefit of anti-AIDS drugs: M. A. Fischl, D. D. Richman, M. H. Grieco, et al., "The Efficacy of Azidothymidine (AZT) in the Treatment of Patients with AIDS and AIDS-Related Complex," *New England Journal of Medicine* 317 (1987): 185–91.

242　Harvard study of asthmatics: M. E. Wechsler, J. M. Kelley, I. O. E. Boyd, et al., "Active Albuterol or Placebo, Sham Acupuncture, or No Intervention in Asthma," *New England Journal of Medicine* 365 (2011): 119–26.

242　Homeopathy and asthma death: Dr. Robert Baratz, personal communication.

242　Linda Epping: Barrett and Jarvis, *Health Robbers*, 83–84.

243　Harm from unpasteurized products: Food and Drug Administration, "The Dangers of Raw Milk: Unpasteurized Milk Can Pose Serious Health Risk," www.fda.gov/Food/ResourcesForYou/Consumers/ucm079516.htm.

243　Mercola and Jumbotron: B. Smith, "Dr. Mercola: Visionary or Quack?" *Chicago*, February 2012.

243　Mercola doesn't believe HIV causes AIDS: Ibid.

243　Mercola, coffee enemas, baking soda: "The Cancer Treatment So Successful—Traditional Doctors Shut It Down," April 23, 2001, http://articles.mercola.com/sites/articles/archive/2011/04/23/dr-nicholas-gonzalez-on-alternative-cancer-treatments.aspx; D. Gorski, "For Shame, Dr. Oz, for Promoting Joseph Mercola on Your Show," *Science-Based Medicine* blog, www.sciencebasedmedicine.org/index.php/for-shame-dr-oz;

Orac, "Dr. Oz: America's Doctor and the Abdication of Professional Responsibility," *ScienceBlogs*, April 13, 2010, http://scienceblogs.com/insolence/2010/04/dr_oz_americas_doctor_and_the_abdication.php; T. Tsouderos, "Questioning Dr. Oz," *Chicago Tribune*, April 9, 2010.

243   Mercola and FDA: T. Tsouderos, "FDA Warns Doctor: Stop Touting Camera As Disease Screening Tool," *Chicago Tribune*, April 25, 2011.

243   Mercola regarding his medical model: "The Alternative Health Controversy," *The Dr. Oz Show*, www.doctoroz.com/videos/alternative-health-controversy-pt.1.

244   Andrew Weil regarding kava: Weil, *Spontaneous Healing*, 321.

244   FDA warning against kava: Food and Drug Administration, "Consumer Advisory: Kava-Containing Dietary Supplements May Be Associated with Severe Liver Injury," March 25, 2002.

244   Baby dies from asphyxiation: Butler, *Consumer's Guide*, 82–83.

244   Neural Organization Technique: Ibid., 171–72.

244   Ernst study regarding harm from chiropractic: E. Ernst, "Deaths After Chiropractic: A Review of Published Cases," *International Journal of Clinical Practice* 64 (2010): 1162–65.

245   Weil and Drugstore.com: Hurley, *Natural Causes*, 240.

245   Chopra's business model: Barrett and Jarvis, *Health Robbers*, 243–45; Butler, *Consumer's Guide*, 118; Johnston, *Politics of Healing*, 223; F. Wheen, *How Mumbo Jumbo Conquered the World: A Short History of Modern Delusions* (New York: Public Affairs, 2004), 44–48.

246   Chopra regarding unlimited wealth: Chopra, *Spiritual Laws*, 1.

246   Chopra regarding money as energy: Ibid., 28.

246   Mercola's business model: D. Gumpert, "Old Time Sales Tricks on the Net," *BusinessWeek*, May 23, 2006.

247   Mercola products: Smith, "Dr. Mercola: Visionary or Quack?"

248   Mercola and government conspiracy: Ibid.

248   Mercola regarding natural products: "The Alternative Health Controversy," *The Dr. Oz Show*.

248   Caplan regarding medical professionalism: Art Caplan, interviewed by the author, September 20, 2011.

249   Titanium necklaces: L. Jenkins, "Is Your Bat Speed a Bit Off? Try a Titanium Necklace," *New York Times*, June 22, 2005.

250   Caplan regarding magical thinking: Art Caplan, interviewed by the author, September 20, 2011.

250 Slack regarding magical thinking: R. Slack, "Acupuncture: A Science-Based Assessment," Center for Inquiry, 2010.

EPILOGUE: ALBERT SCHWEITZER AND THE WITCH DOCTOR

254 N. Cousins and A. Schweitzer: Cousins, *Anatomy of an Illness*, 76–78.

Selected Bibliography

Alcabes, Philip. *Dread: How Fear and Fantasy Have Fueled Epidemics from the Black Death to Avian Flu.* New York: Public Affairs, 2009.

Anderson, Ann. *Snake Oil, Hustlers and Hambones: The American Medicine Show.* Jefferson, N.C.: McFarland & Company, 2000.

Anonymous. *Nostrums and Quackery: Articles on the Nostrum Evil and Quackery Reprinted, with Additions and Modifications, from the Journal of the American Medical Association.* Chicago: American Medical Association Press, 1912.

Barrett, Stephen. *Health Schemes, Scams, and Frauds.* Mount Vernon, N.Y.: Consumer Reports Books, 1990.

Barrett, Stephen, and William T. Jarvis, eds. *The Health Robbers: A Close Look at Quackery in America.* Buffalo: Prometheus Books, 1993.

Barrett, Stephen, and Victor Herbert. *The Vitamin Pushers: How the "Health Food" Industry Is Selling America a Bill of Goods.* Amherst, N.Y.: Prometheus Books, 1994.

Barrett, Stephen, William London, Robert Bartz, and Manfred Kroger, eds. *Consumer Health: A Guide to Intelligent Decisions.* New York: McGraw-Hill, 2007.

Bausell, R. Barker. *Snake Oil Science: The Truth About Complementary and Alternative Medicine.* Oxford: Oxford University Press, 2007.

Bean, Constance A., and Lesley A. Fein. *Beating Lyme: Understanding and Treating This Complex and Often Misdiagnosed Disease.* New York: AMACOM, 2008.

Benedetti, Fabrizio. *Placebo Effects: Understanding the Mechanisms in Health and Disease.* Oxford: Oxford University Press, 2009.

Benedetti, Paul, and Wayne MacPhail. *Spin Doctors: The Chiropractic Industry Under Examination.* Toronto: Dundurn Press, 2002.

Brock, Pope. *Charlatan: America's Most Dangerous Huckster, the Man Who Pursued Him, and the Age of Flimflam.* New York: Three Rivers Press, 2008.

Buhner, Steven H. *Healing Lyme: Natural Healing and Prevention of Lyme Borreliosis and Its Coinfections.* Silver City, N.M.: Raven Press, 2005.

Butler, Kurt. *A Consumer's Guide to "Alternative Medicine": A Close Look at Homeopathy, Acupuncture, Faith-Healing, and Other Unconventional Treatments.* Amherst, N.Y.: Prometheus Books, 1992.

Buttar, Rashid A. *The 9 Steps to Keep the Doctor Away: Simple Actions to Shift the Body and Mind to Optimal Health for Greater Longevity.* Lake Tahoe, Nev: GMEC Publishing, 2010.

Carroll, Robert Todd. *The Skeptic's Dictionary: A Collection of Strange Beliefs, Amusing Deceptions & Dangerous Delusions.* Hoboken, N.J.: John Wiley & Sons, 2003.

Chopra, Deepak. *Quantum Healing: Exploring the Frontiers of Mind/Body Medicine.* New York: Bantam Books, 1989.

———. *Perfect Weight: The Complete Mind/Body Program for Achieving and Maintaining Your Ideal Weight.* New York: Three Rivers Press, 1991.

———. *Ageless Body, Timeless Mind: The Quantum Alternative to Growing Old.* New York: Three Rivers Press, 1993.

———. *Creating Affluence: The A-to-Z Steps to a Richer Life.* San Rafael, Calif.: Amber-Allen Publishing, 1993.

———. *The Seven Spiritual Laws of Success: A Practical Guide to the Fulfillment of Your Dreams.* San Rafael, Calif.: Amber-Allen Publishing, 1994.

———. *The Path to Love: Spiritual Strategies for Healing.* New York: Three Rivers Press, 1997.

———. *The Book of Secrets: Unlocking the Hidden Dimensions of Your Life.* New York: Three Rivers Press, 2004.

———. *Reinventing the Body, Resurrecting the Soul: How to Create a New You.* New York: Three Rivers Press, 2009.

Chopra, Deepak, and David Simon. *Grow Younger, Live Longer: 10 Steps to Reverse Aging.* New York: Three Rivers Press, 2001.

Clark, Hulda Regehr. *The Cure for All Cancers.* Chula Vista, Calif.: New Century Press, 1993.

———. *The Cure for HIV and AIDS.* Chula Vista, Calif.: New Century Press, 1993.

———. *The Cure for All Diseases.* Chula Vista, Calif.: New Century Press, 1995.

Cousins, Norman. *Anatomy of an Illness.* New York: W. W. Norton & Company, 1979.

———. *Head First: The Biology of Hope and the Healing Power of the Human Spirit.* New York: Penguin Books, 1989.

Cramp, Arthur, ed. *Nostrums and Quackery: Articles on the Nostrum Evil, Quackery and Allied Matters Affecting the Public Health: Reprinted, With or Without Modifications, from the Journal of the American Medical Association*, vol. II. Chicago: American Medical Association Press, 1921.

Diamond, John. *Snake Oil and Other Preoccupations*. London: Vintage, 2001.

Elias, Thomas D. *The Burzynski Breakthrough*. Nevada City, Calif.: Lexikos, 2001.

Ernst, Edzard, ed. *Healing, Hype, or Harm? A Critical Analysis of Complementary or Alternative Medicine*. Charlottesville, Va.: Societas, 2008.

Evans, R. L., and I. M. Berent. *Drug Legalization: For and Against*. La Salle, Ill.: Open Court, 1992.

Fishbein, Morris. *The Medical Follies*. New York: Boni and Liveright, 1925.

———. *The New Medical Follies*. New York: Boni and Liveright, 1927.

———. *Fads and Quackery in Healing: An Analysis of the Foibles of the Healing Cults, with Essays on Various Other Peculiar Notions in the Health Field*. New York: Blue Ribbon Books, 1932.

Fonorow, Owen. *Practicing Medicine Without a License? The Story of the Linus Pauling Therapy for Heart Disease*. Lulu.com, 2008.

Food Protection Committee, Food and Nutrition Board, National Academy of Sciences, National Research Council. *Toxicants Occurring Naturally in Foods*. Washington, D. C.: National Academy of Sciences Press, 1966.

Frazier, Kendrick, ed. *Science Under Siege: Defending Science, Exposing Pseudoscience*. Amherst, N.Y.: Prometheus Books, 2009.

Goertzel, Ted, and Ben Goertzel. *Linus Pauling: A Life in Science and Politics*. New York: Basic Books, 1995.

Goldacre, Ben. *Bad Science*. London: HarperCollins, 2008.

Hager, Thomas. *Force of Nature: The Life of Linus Pauling*. New York: Simon & Schuster, 1995.

———. *Linus Pauling and the Chemistry of Life*. Oxford: Oxford University Press, 1998.

Harrington, Anne, ed. *The Placebo Effect: An Interdisciplinary Exploration*. Cambridge, Mass.: Harvard University Press, 1997.

Hawkins, David, and Linus Pauling. *Orthomolecular Psychiatry: Treatment of Schizophrenia*. San Francisco: W. H. Freeman and Company, 1973.

Helfand, William H. *Quack, Quack, Quack: The Sellers of Nostrums in Prints, Posters, Ephemera & Books*. New York: Grolier Club, 2002.

Herbert, Victor. *Nutrition Cultism: Facts and Fictions*. Philadelphia: George F. Stickley Company, 1980.

Herbert, Victor, and Stephen Barrett. *Vitamins and "Health" Foods: The Great American Hustle*. Philadelphia: George F. Stickley Company, 1981.

Hoffer, Abram, and Linus Pauling. *Healing Cancer: Complementary Vitamin and Drug Treatments*. Toronto: CCNM Press, 2004.

Hood, Bruce. *The Science of Superstition: How the Developing Brain Creates Supernatural Beliefs*. New York: HarperCollins, 2009.

Hurley, Dan. *Natural Causes: Death, Lies, and Politics in America's Vitamin and Herbal Supplement Industry*. New York: Broadway Books, 2006.

Jacoby, Susan. *Never Say Die: The Myth and Marketing of the New Old Age*. New York: Pantheon Books, 2011.

Jameson, Eric. *The Natural History of Quackery*. Springfield, Ill.: Charles C. Thomas, 1961.

Jepson, Bryan. *Changing the Course of Autism: A Scientific Approach for Parents and Physicians*. Boulder, Colo.: Sentient Publishing, 2007.

Jesson, Lucinda, and Stacey Tovino. *Complementary and Alternative Medicine and the Law*. Durham, N.C.: Carolina Academic Press, 2010.

Johnston, Robert, ed. *The Politics of Healing: Histories of Alternative Medicine in Twentieth-Century North America*. New York: Routledge, 2004.

Juhne, Eric S. *Quacks and Crusaders: The Fabulous Careers of John Brinkley, Norman Baker, & Harry Hoxsey*. Lawrence: University Press of Kansas, 2002.

Kabat, Geoffrey C. *Hyping Health Risks: Environmental Hazards in Daily Life and the Science of Epidemiology*. New York: Columbia University Press, 2011.

Kalichman, Seth. *Denying AIDS: Conspiracy Theories, Pseudoscience, and Human Tragedy*. New York: Copernicus Books, 2009.

Kennedy, Dan S., and Chip Kessler. *Making Them Believe: How One of America's Legendary Rogues Marketed "The Goat Testicles Solution" and Made Millions*. Garden City, N.Y.: Glazer-Kennedy Publishing, 2010.

Kradin, Richard. *The Placebo Response and the Power of Unconscious Healing*. New York: Routledge, 2008.

Lee, R. Alton. *The Bizarre Careers of John R. Brinkley*. Lexington: The University Press of Kentucky, 2002.

Lefkowitz, L. J., Attorney General, State of New York, by D. K. McGivney, Esq., Appendix on Appeal, *In the Matter of Joseph Hofbauer*, State of New York Supreme Court, Appellate Division, Third Judicial Department, Index No. N-46-1164-77, May 17, 1978, A1–A1610.

Lerner, Barron. *When Illness Goes Public: Celebrity Patients and How We Look at Medicine*. Baltimore: Johns Hopkins University Press, 2006.

Marinacci, Barbara, ed. *Linus Pauling in His Own Words: Selections from His Writings, Speeches, and Interviews*. New York: Simon & Schuster, 1995.

McCarthy, Jenny. *Mother Warriors: A Nation of Parents Healing Autism Against All Odds*. New York: Plume, 2008.

McCarthy, Jenny, and Jerry Kartzinel. *Healing and Preventing Autism: A Complete Guide*. New York: Plume, 2010.

McCoy, Bob. *Quack! Tales of Medical Fraud from the Museum of Questionable Medical Devices*. Santa Monica, Calif.: Santa Monica Press, 2000.

McFadzean, Nicola. *The Lyme Diet: Nutritional Strategies for Healing Lyme Disease*. San Diego: Legacy Line Publishing, 2010.

McNamara, Brooks. *Step Right Up*. Jackson: University of Mississippi Press, 1995.

Mead, Clifford, and Thomas Hager, eds. *Linus Pauling: Scientist and Peacemaker*. Corvallis: Oregon State University Press, 2001.

Modde, Peter J. *Chiropractic Malpractice*. Columbia, Md.: Henrow Press, 1985.

Mooney, Chris, and Sheril Kirshenbaum. *Unscientific America: How Scientific Illiteracy Threatens Our Future*. New York: Basic Books, 2009.

Newton, David E. *Linus Pauling: Scientist and Advocate*. New York: Facts on File, 1994.

Northrup, Christiane. *The Wisdom of Menopause: Creating Physical and Emotional Health During the Change*. New York: Bantam Dell, 2006.

———. *Women's Bodies, Women's Wisdom: Creating Physical and Emotional Health and Healing*. New York: Bantam Books, 2010.

Offit, Paul. *Autism's False Prophets: Bad Science, Risky Medicine, and the Search for a Cure*. New York: Columbia University Press, 2008.

———. *Deadly Choices: How the Anti-Vaccine Movement Threatens Us All*. New York: Basic Books, 2011.

Park, Robert. *Voodoo Science: The Road from Foolishness to Fraud*. Oxford: Oxford University Press, 2000.

———. *Superstition: Belief in the Age of Science*. Princeton, N.J.: Princeton University Press, 2008.

Pauling, Linus. *Vitamin C and the Common Cold*. San Francisco: W. H. Freeman and Company, 1970.

———. *Vitamin C, the Common Cold, and the Flu*. San Francisco: W. H. Freeman and Company, 1976.

———. *How to Live Longer and Feel Better*. Corvallis: Oregon State University Press, 1986.

Pauling, Linus, and Ewan Cameron, eds. *Cancer and Vitamin C: A Discussion of the*

Nature, Causes, Prevention, and Treatment of Cancer with Special Reference to the Value of Vitamin C. Philadelphia: Camino Books, 1979 (updated 1993).

Perper, Joshua, and Stephen Cina. *When Doctors Kill: Who, Why, and How*. New York: Copernicus Books, 2010.

Piazza, Gail, and Laura Piazza. *Recipes for Repair: A Lyme Disease Cookbook*. Sunapee, N.H.: Peconic Publishing, 2010.

Pierce, Charles. *Idiot America: How Stupidity Became a Virtue in the Land of the Free*. New York: Anchor Books, 2009.

Plotkin, Stanley A., Walter A. Orenstein, and Paul A. Offit, eds. *Vaccines*, 6th ed. Philadelphia: Saunders, 2012.

Porter, Roy. *Quacks: Fakers & Charlatans in Medicine*. Gloucestershire, England: Tempus, 2003.

Randi, James. *Flim-Flam: Psychics, ESP, Unicorns and Other Delusions*. Amherst, N.Y.: Prometheus Books, 1982.

———. *The Faith Healers*. Amherst, N.Y.: Prometheus Books, 1989.

———. *An Encyclopedia of Claims, Frauds, and Hoaxes of the Occult and Supernatural*. New York: St. Martin's Griffin, 1995.

Roizen, Michael F., and Mehmet C. Oz. *You, the Owner's Manual: An Insider's Guide to the Body That Will Make You Healthier and Younger*. New York: Collins, 2005.

———. *You Staying Young: The Owner's Manual for Extending Your Warranty*. New York: Free Press, 2007.

———. *You Being Beautiful: The Owner's Manual to Inner and Outer Beauty*. New York: Free Press, 2008.

———. *You Having a Baby: The Owner's Manual to a Happy and Healthy Pregnancy*. New York: Free Press, 2009.

———. *You Raising Your Child: The Owner's Manual from First Breath to First Grade*. New York: Free Press, 2010.

Rosner, Bryan. *The Top 10 Lyme Disease Treatments: Defeat Lyme Disease with the Best of Conventional and Alternative Medicine*. South Lake Tahoe, Calif.: BioMed Publishing Group, 2007.

———. *Lyme Disease and Rife Machines*. South Lake Tahoe, Calif.: BioMed Publishing Group, 2005.

Schwarcz, Joe. *Radar, Hula Hoops, and Playful Pigs: 67 Digestible Commentaries on the Fascinating Chemistry of Everyday Food & Life*. Toronto: ECW Press, 1999.

———. *The Fly in the Ointment: 70 Fascinating Commentaries on the Science of Everyday Food & Life*. Toronto: ECW Press, 2004.

———. *Let Them Eat Flax: 70 All-New Commentaries on the Science of Everyday Food & Life.* Toronto: ECW Press, 2005.

———. *Brain Fuel: 199 Mind-Expanding Inquiries into the Science of Everyday Life.* Scarborough, Ontario: Doubleday Canada, 2008.

———. *Science, Sense and Nonsense: 61 Nourishing, Healthy, Bunk-Free Commentaries on the Chemistry That Affects Us All.* Scarborough, Ontario: Doubleday Canada, 2009.

Serafini, Anthony. *Linus Pauling: A Man and His Science.* Lincoln, Neb.: ToExcel Press, 1989.

Shapiro, Arthur, and Elaine Shapiro. *The Powerful Placebo: From Ancient Priest to Modern Physician.* Baltimore: Johns Hopkins University Press, 1997.

Shapiro, Rose. *Suckers: How Alternative Medicine Makes Fools of Us All.* London: Harvill Secker, 2008.

Sherrow, Victoria. *Linus Pauling: Investigating the Magic Within.* Austin: Raintree Steck-Vaughn Publishers, 1997.

Silver, Lee M. *Challenging Nature: The Clash Between Biotechnology and Spirituality.* New York: Ecco, 2006.

Singh, Simon, and Edzard Ernst. *Trick or Treatment: The Undeniable Facts About Alternative Medicine.* New York: W. W. Norton, 2008.

Singleton, Kenneth B. *The Lyme Disease Solution.* Charleston, S.C.: BookSurge Publishing, 2008.

Siri, Ken, and Tony Lyons. *Cutting-Edge Therapies for Autism: 2010-2011.* New York: Skyhorse Publishing, 2010.

Smith, Ralph Lee. *The Health Hucksters.* New York: Thomas Y. Crowell Co., 1960.

Somers, Suzanne. *The Sexy Years: Discover the Hormone Connection: The Secret to Fabulous Sex, Great Health, and Vitality, for Women and Men.* New York: Three Rivers Press, 2004.

———. *Slim & Sexy Forever: The Hormone Solution for Permanent Weight Loss and Optimal Living.* New York: Three Rivers Press, 2005.

———. *Ageless: The Naked Truth About Bioidentical Hormones.* New York: Three Rivers Press, 2006.

———. *Breakthrough: Eight Steps to Wellness: Life-Altering Secrets from Today's Cutting-Edge Doctors.* New York: Three Rivers Press, 2008.

———. *Knockout: Interviews with Doctors Who Are Curing Cancer and How to Prevent Getting It in the First Place.* New York: Crown Publishing Group, 2009.

———. *Sexy Forever: How to Fight Fat After Forty: Shed the Toxins, Shed the Fat.* New York: Crown Archetype, 2010.

Specter, Michael. *Denialism: How Irrational Thinking Hinders Scientific Progress, Harms the Planet, and Threatens Our Lives*. New York: Penguin Press, 2009.

Starr, Paul. *The Social Transformation of American Medicine*. New York: Basic Books, 1982.

Stoddard, George D. *"Krebiozen": The Great Cancer Mystery*. Boston: Beacon Press, 1955.

Stone, Irwin. *The Healing Factor: Vitamin C Against Disease*. New York: Grosset & Dunlap, 1972.

Storl, Wolf D. *Healing Lyme Disease Naturally: History, Analysis, and Treatments*. Berkeley, Calif.: North Atlantic Books, 2010.

Strasheim, Connie. *Insights into Lyme Disease Treatment: 13 Lyme Literate Health Care Practitioners Share Their Healing Strategies*. South Lake Tahoe, Calif.: BioMed Publishing Group, 2009.

Tallis, Raymond. *Hippocratic Oaths: Medicine and Its Discontents*. London: Atlantic Books, 2005.

Thompson, W. Grant. *The Placebo Effect and Health: Combining Science and Compassionate Care*. Amherst, N.Y.: Prometheus Books, 2005.

Wanjek, Christopher. *Bad Medicine: Misconceptions and Misuses Revealed, from Distance Healing to Vitamin O*. Hoboken, N.J.: John Wiley & Sons, 2003.

Weil, Andrew. *Health and Healing*. Boston: Houghton Mifflin, 1983.

——. *Natural Health, Natural Medicine: The Complete Guide to Wellness and Self-Care for Optimum Health*. Boston: Houghton Mifflin, 1995.

——. *Spontaneous Healing: How to Discover and Enhance Your Body's Natural Ability to Maintain and Heal Itself*. New York: Balantine Books, 1995.

——. *8 Weeks to Optimum Health: A Proven Program for Taking Full Advantage of Your Body's Natural Healing Power*. New York: Alfred A. Knopf, 1997.

——. *Healthy Aging: A Lifelong Guide to Your Well-Being*. New York: Anchor Books, 2005.

——. *You Can't Afford to Get Sick: Your Guide to Optimum Health and Health Care*. New York: Plume, 2009.

Wheen, Francis. *How Mumbo Jumbo Conquered the World: A Short History of Modern Delusions*. New York: Public Affairs, 2004.

White, Florence Meiman. *Linus Pauling: Scientist and Crusader*. New York: Walker and Company, 1980.

Wright, Jonathan V., and Lane Lenard. *Stay Young and Sexy with Bioidentical Hormone Replacement: The Science Explained*. Petaluma, Calif.: Smart Publications, 2010.

Young, John Harvey. *The Medical Messiahs: A Social History of Health Quackery in Twentieth-Century America*. Princeton, N.J.: Princeton University Press, 1967.

———. *The Toadstool Millionaires: A Social History of Patent Medicines in America Before Federal Regulation*. Princeton, N.J.: Princeton University Press, 1972.

———. *American Health Quackery*. Princeton, N.J.: Princeton University Press, 1992.

Index

About the Author

P aul A. Offit, M.D., is chief of the Division of Infectious Diseases and director of the Vaccine Education Center at the Children's Hospital of Philadelphia, as well as the acclaimed author of *Autism's False Prophets*, *Vaccinated*, and *Deadly Choices*.

www.paul-offit.com

BOOKS BY PAUL A. OFFIT, MD

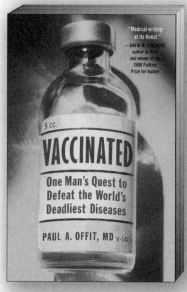

VACCINATED
One Man's Quest to Defeat the World's Deadliest Diseases

Available in Paperback and eBook

Maurice Hilleman became a microbiologist and joined Merck, the pharmaceutical company, to pursue his goal of eliminating childhood disease. Chief among his accomplishments are nine vaccines that practically every child gets, rendering formerly dread diseases—including mumps and rubella—practically toothless and nearly forgotten; his measles vaccine alone saves several million lives every year. *Vaccinated* is not a biography; Hilleman's experience forms the basis for a rich and lively narrative of two hundred years of medical history. It is an inspiring and triumphant tale, but one with a cautionary aspect, as vaccines come under assault from people blaming them for autism and worse. Paul Offit rebuts those arguments, and, by demonstrating how much the work of Hilleman and others has gained for humanity, shows us how much we have to lose.

DO YOU BELIEVE IN MAGIC?
Vitamins, Supplements, and All Things Natural: A Look Behind the Curtain

Available in Paperback and eBook

A half century ago, acupuncture, homeopathy, naturopathy, Chinese herbs, Christian exorcisms, dietary supplements, chiropractic manipulations, and ayurvedic remedies were considered on the fringe of medicine. Now these practices have become mainstream, used by half of all Americans today to treat conditions from losing weight to preventing cancer. But as Offit reveals, many popular alternative therapies are ineffective, expensive, and even deadly. He debunks the treatments that don't work and explains why, and takes on the media celebrities who promote alternative medicine. Using real-life stories, he separates the sense from the nonsense, showing why any therapy should be scrutinized and showing how some of these remedies really do work.